Springer Tracts in Civil Engineering

Springer Tracts in Civil Engineering (STCE) publishes the latest developments in Civil Engineering - quickly, informally and in top quality. The series scope includes monographs, professional books, graduate textbooks and edited volumes, as well as outstanding PhD theses. Its goal is to cover all the main branches of civil engineering, both theoretical and applied, including:

- Construction and Structural Mechanics
- Building Materials
- Concrete, Steel and Timber Structures
- Geotechnical Engineering
- Earthquake Engineering
- Coastal Engineering; Ocean and Offshore Engineering
- Hydraulics, Hydrology and Water Resources Engineering
- Environmental Engineering and Sustainability
- Structural Health and Monitoring
- Surveying and Geographical Information Systems
- Heating, Ventilation and Air Conditioning (HVAC)
- Transportation and Traffic
- Risk Analysis
- Safety and Security

Indexed by Scopus

To submit a proposal or request further information, please contact:
Pierpaolo Riva at Pierpaolo.Riva@springer.com (Europe and Americas) Wayne Hu at wayne.hu@springer.com (China)

More information about this series at http://www.springer.com/series/15088

Omoseni Adepoju · Clinton Aigbavboa ·
Nnamdi Nwulu · Michael Onyia

Re-skilling Human Resources for Construction 4.0

Implications for Industry, Academia and Government

Springer

Omoseni Adepoju
Faculty of Engineering and the Built Environment
University of Johannesburg
Johannesburg, South Africa

Clinton Aigbavboa
Faculty of Engineering and the Built Environment
University of Johannesburg
Johannesburg, South Africa

Nnamdi Nwulu
Department of Electrical and Electronic Engineering Science
University of Johannesburg
Johannesburg, South Africa

Michael Onyia
Department of Civil Engineering
University of Nigeria
Nsukka, Nigeria

ISSN 2366-259X ISSN 2366-2603 (electronic)
Springer Tracts in Civil Engineering
ISBN 978-3-030-85972-5 ISBN 978-3-030-85973-2 (eBook)
https://doi.org/10.1007/978-3-030-85973-2

This Springer imprint is published by the registered company Springer Nature Switzerland AG
The registered company address is: Gewerbestrasse 11, 6330 Cham, Switzerland

Preface

The fourth industrial revolution, also known as Industry 4.0, has transformed the construction industry by creating digital, smart and sustainable construction technologies. The innovative deployment of Industry 4.0 concepts and its application in the construction industry is called Construction 4.0. These new technologies such as Building Information Modelling (BIM), Robotics, 3D printing and Drones give the professionals in the industry a different outlook about optimising their operations to boost the delivery of cost-effective, timely and quality projects. The application of these technologies comes with several benefits and some risks. More importantly, the adoption of these technologies demand new skills for its operation and effective performance.

In order to realise the benefits of these modern technologies, the construction industry needs to address the inevitable challenges, especially those pertaining to human resources. The availability of required skills will improve the implementation of Construction 4.0 technologies. Consequently, it becomes necessary to develop and build talents for the present and future needs of the industry. In line with this, the stakeholders in the industry have roles to play in building the necessary capabilities required to rapidly adapt to the new technologies. The industry is responsible for training the workers, creating a learning environment for practical application of the new technologies and working closely with educational institutions to equip personnel with the required knowledge and skills for Construction 4.0. Also, educational and training institutions with the collaboration of the industry are required to design courses and training programmes that can help nurture, develop and re-skill the human resources suited for a digitalised construction industry.

Furthermore, the governments have a responsibility in policy formulation to reinforce human capital development both at the educational institutions and the workplace. Hence, this research book will examine the key technologies of Construction 4.0, the benefits, risks and relevant skills required to implement the technologies. The book will also consider human resources reskilling needs and the roles of key stakeholders in developing the skills needed for Construction 4.0. This research views the lack of twenty-first century skills and the skills gap, especially in developing countries, as a limiting factor to implementing these technologies.

Johannesburg, South Africa	Omoseni Adepoju
Johannesburg, South Africa	Clinton Aigbavboa
Johannesburg, South Africa	Nnamdi Nwulu
Nsukka, Nigeria	Michael Onyia

Contents

Part I
Introduction and Background to Reskilling and Upskilling for Construction 4.0 Technologies

Chapter 1
Introduction

1.1 Background to the Study

The construction industry, a significant driver of economic growth of nations, is currently experiencing transformation through advancements in technology ushered by the fourth industrial revolution, also known as Industry 4.0. This transformation is demanding new approaches and specific skills requirements for effective implementation. This revolution is significant as the construction industry represents an economic booster and a major source of employment opportunities. This is also reflected in the 2018 analysis of McKinsey, which reveals that global spending in the construction industry was $11 trillion, with a projection to $814 trillion in 2025, accounting for 9% of World GDP. In sub-Saharan Africa, the construction industry market is projected to register a 6.4% CAGR by the end of 2024 (Africa Review, 2021). An observation of the built environment and humans' daily activities reveals that the construction industry shapes humanity's economic, social, and environmental aspects.

Industry 4.0 is premised on combining physical and digital technologies with data, cyber-physical systems and the Internet of things (IoT) in ensuring smart production and processes. The cyber-physical system, which is the core of the fourth industrial revolution, comprises a pair of technologies that connect the physical and virtual worlds. These technologies can create an intelligent environment where smart objects communicate and interact with one another. The technologies of industry 4.0 include the Internet of Things (IoT), Big Data, Augmented and Virtual Reality, Simulation, cybersecurity, Cloud computing, Robotics, Prefabrication/Additive manufacturing and integration of Horizontal and Vertical systems. These nine (9) pillars have reshaped, reprogrammed and refocused the process, technologies, and methodologies, of many industries, such as the Construction industry. Industries can be assured of operational efficiency, optimised profitability, better quality, increased productivity, customer and stakeholder satisfaction, and more innovations.

Industry 4.0 technologies exert influence and pressure on other industries, especially the construction industry, which is the fulcrum of humanity. The influence of

O. Adepoju et al., *Re-skilling Human Resources for Construction 4.0*, Springer Tracts in Civil Engineering, https://doi.org/10.1007/978-3-030-85973-2_1

industry 4.0 technologies on the construction industry is known as construction 4.0. The construction industry Fourth Industrial Revolution is the application of industry 4.0 in the construction industry (Adepoju & Aigbavboa, 2020). Construction 4.0, according to Sawhney et al. (2020), is "*a paradigm that uses cyber-physical systems, and the internet of things, data and services to link the digital layer consisting of BIM and Common Data Environment (CDE) and the physical layer consisting of the asset over its whole life to create an interconnected environment integrating organisations, process, and information to efficiently design, construct and operate assets*". Construction 4.0, as adapted from industry 4.0 technologies, is the application of emerging technologies, digital processes and cyber-physical procedures to construction activities in the industry. The Fourth industrial revolution in the construction industry is a new dawn of processes driven by digital technologies permeating the architectural, engineering, construction and facility management of the construction industry.

Moreover, construction 4.0 operates on four (4) design principles of interconnection and interoperability, technical assistance (seen in the usage of drones, robots, 3D printing), Information transparency (occasioned by virtual reality and augmented reality in construction processes) and decentralised decision making (Hossain & Nadeem, 2019). Hence, construction 4.0 brings about digital clarity in construction procedures and automated processes in construction activities. The application of Industry 4.0 to Construction can be summarised under six principles which include: virtualisation, decentralisation, modularity, service orientation, interoperability and real-time capability (Hermann, Pentek, & Otto, 2016). Virtualisation is the ability of devices with the use of sensors to monitor physical procedures and communication between machines. The principle of decentralisation entails decentralised decision making empowered through computer systems installed for devices, workers and organisations. Modularity entails establishing a modular manufacturing system, which creates manufactured components that can be assembled, expanded, flexible, and replaced. The principle of service orientation emphasised that the technologies must solve problems and satisfy needs and wants. Interoperability is the ability of technologies to connect, communicate with each other and comprehend the functionality of one another. Real-time capability entails the immediate collection, transmission and analysis of data required by machines and men for effective operations and decision-making processes.

All these principles and models of the fourth industrial revolution have shaped both the construction industry, demanding an adjustment to new ways of building construction projects, new ways of living inside construction projects, and new ways of maintaining finished projects.

1.1.1 Construction 4.0

Construction 4.0 revolves around creating a digital construction site through internet-connected sensors on equipment and its application at modern technologies at every stage of the construction project (Osunsanmi, Aigbavboa & Oke, 2018). This concept has introduced several changes in the construction industry, such as the application of industry 4.0 technologies, mass production, green Construction, increase in professional technological proficiency and customisation. The Construction 4.0 technologies include: Building information modelling (BIM), Big data, Internet of things and services, unmanned aerial vehicle (UAV), additive manufacturing (3D printing), modular prefabricated Construction, 3D scanner, cloud computing, Augmented reality (AR)/ Virtual reality (VR)/ Mixed Reality (MR), simulations of virtual robotics/ autonomous vehicle, GPS, Artificial intelligence, sensors and actuators, cyber-physical systems and Radio-Frequency Identification (RFID) (Kozlovska, Klosova & Strukova, 2021). Most of this technology is infused in the construction industry during application in construction activities. The most prominent Construction 4.0 technologies are Building Information Modeling (BIM), Drones and Unmanned aerial vehicles (UAC), prefabrication, 3D printings, Robotic Construction and the Internet of things. The effective implementation and utilisation of these technologies require specific skills and a technologically oriented environment. However, this book explored six major Construction 4.0 technologies in the construction industry. These are discussed below:

1.1.2 Building Information Modeling (BIM)

The construction industry optimises the construction data and information through building information modelling (BIM). According to Mohandes and Omrany (2013), the BIM is a project simulation technology consisting of three dimensional (3D) models of project components by connecting all the project data and information to the project planning, project constructing and decommissioning. The US National Institute of Building Sciences (2007) defined the BIM as "*a digital representation of physical and functional characteristics of a facility and a shared knowledge resource for information about a facility forming a reliable basis for decision making during its life cycle, defined as existing from earliest conception to demolition*". According to Tahir et al. (2017), the BIM technology applications in the construction industry includes design and visualisation, scheduling, quantity take-off and estimation, quality and risk control, integration and collaboration of stakeholders, clash detection and coordination, analysis of constructability, material resource control, human resource management, facility management, reduction of reworking during Construction, contract administration and efficient communication. The BIM is an integrative technology that collects and use construction stages. Mohandes and Omramy (2013) opined that BIM is very important to the construction industry due to its numerous benefits, such as improvement in project visualisation, proper team management, analysis of complex details, effective project scheduling, estimations of work breakdown structure, quick site preparations, enabling the best routes for pipes, lights, cables, wires, ductwork and sprinklers; proper scheduling for lifts for the utilisation of concrete, steel, electrical equipment placement and huge mechanical equipment placement, evaluation of needed safety precautions and adequate coordination of construction activities. This unequivocally improves construction works, site effectiveness, waste reduction and management, cost-effectiveness and high project performance. The BIM technology is built with various software, including Revit, Revizto, Navisworks, ArchiCAD, Vectorworks Architect, Edificius, Autodesk BIM 360, Sketchup, Buildertrend, BIMobject, Civil 3D, BricsCAD BIM, Sefaria, VisualARQ, AllPlan Architecture, ActCAD BIM (Ocean, 2020).

1.1.3 Drones and Unmanned Aerial Vehicles

Construction works are often complex, depending on the project size or nature of the project site. However, for efficiency, drones which happen to be one of the industry 4.0 technologies, can be utilised. Drones, also known as unmanned aerial vehicles (UAVs) or unmanned aircraft systems (UAS), are like an aircraft without anybody controlled from a remote place. Drones are equipped with downward–facing sensors such as LIDAR, RGB for capturing aerial data in a short time (Wingtra, 2021). Tkac and Mesaros (2019) opined that the utilisation of drones in the construction industry had served many purposes, which includes: inspection of highways, bridges, cell towers, wind turbines, high mast lighting, building façade, surveying and mapping, drainage and erosion, traffic monitoring and wetland. Drone usage in construction 4.0 is mainly for data collection and monitoring purposes. In collecting and collating data from the built environment and construction sites, drones can access and navigates narrow and dangerous locations. The high-quality data collected through drones have helped in construction management in the following areas of site coordination, safety, dispute resolution, quality control, planning and designing documentation of field conditions, the performance of large scale analysis, time management, cost efficiency, aiding better project performance, building surveys, laser scanning and aerial photogrammetric for scanning buildings, thermal image recording for assessing cold spots in buildings (McCoy & Yeganeh, 2021; Tkac & Mesaros, 2019). According to Tkac and Mesaros (2019) and Chapman (2013), different types of Drones are used for construction works, and these include: multi-rotor drones (primarily used for aerial mapping, aerial photography, aerial surveying and video recording); fixed-wing drones, (high altitude drones, used for efficient for aerial mapping topography); single rotor drones (drones with one rotor but last longer in the air than multiple rotors), and fixed-wing hybrid, vertical takeoff and landing rotor (a drone that flies on a pre-scheduled flight route at user-specified heights and collects data through its colour and multispectral sensors).

1.1.4 *Prefabrication*

Ineffective material management has been a constant headache for construction professionals, as materials are the cornerstones of a construction project. This is why prefabrication technology has been adopted. It entails producing construction components off-sites and then assembled on-site to create/ build the needed structure. According to Li et al. (2014), prefabrication entails producing various construction components at a specialised facility before final assembling and installation. This process is against conventional construction processes of the cast–in–place of concrete, timber formwork, bamboo scaffolding, tilling, reinforcement, and plastering. Auti and Patil (2018) opined that prefabrication technology is the construction process that entails: the finalisation of the design of various components of casting, structure and curing of units, transportation of units on site and the installation of those units. Mays (2021) stated that advantages of prefabricated technology include: a reduction in labour costs, improvement in labour safety, minimisation of construction delays, increase in high quality of work, sminimisation of waste, increases paces of work and increase in energy efficiency.

1.1.5 *3D Printing*

Another construction 4.0 technology is 3D printing, a digital fabrication technology that produces physical object 3D printing. This digital fabrication technology produces physical objects in a dimensional structure of any shape from a geometrical representation (Shahrubudin, Lee, & Ramlan, 2019). Also, Hossain et al. (2020) stated that 3D (three dimensional) printings, which is industrially known as additive manufacturing, are an automated process that produces 3D shapes using a computer-aided design (CAD) on a layer by layer basis to get the required results. The use of 3D printing technology in the construction industry, which is enhanced by other construction software, is more about the precision of buildings and buildings parts coupled with time effectiveness. Also, in 3D printing, different materials are integrated to print a particular building layer, such as binders, cement, sand, aggregate, and water. 3D printing layers are primarily manufactured and produced off-site before being assembled on the construction site. In the construction industry, the 3D printing technology aids the reduction of a manual process, reduce construction waste by 30–60%, construction time by 50–70%, labour costs by 50–80%, material costs by 65%, and it is environmentally friendly (Markets & Markets, 2020; Hossain et al., 2020; Allouzi, Al-Azhain & Allouzi, 2020). 3D printing is enabled by 3D printers, which includes stereolithography (SLA), selective laser sintering (SLS), fused deposition modelling, FDIM (Formlabs, 2021). Other printers, according to Cherdo (2021), includes; BetAbramPi, COBOD BOD 2, 3D constructor, Cyber construction, ICON Vulcan II, MudBots 3D concrete printer, Total Kustom story Bot 6.2, WASP Crane WASP, APIS cor, Bati print 3D, SQ4D—ARCS, Contour crafting and Xtree E. However, before 3D printings, they are designed in several 3D software, depending on construction requirements. These include Sketch-up, AutoCAD, ArchiCAD, AutoCAD Architecture, DataCAD, Revit, Cedreo, Auto CAD Civil 3D, chief Architect, All plan Vectorworks Architect, Microstation, Bricscad BIM, Rhino 3D, CorelCAD, Solid works (Sculpteo, 2021).

1.1.6 Robotic Construction

Consequently, like most industries, data and information are the operational bedrock of the construction industry. This data feeds into different construction processes and technologies to optimise the satisfaction of end-users and maximise shareholders' wealth.

Moreover, construction activities, processes and procedures require excessive human efforts to achieve project performance, which has led many stakeholders and investors to use automation through Robotics for efficiency and effectiveness. Robotic technology is a technology designed to replace and perfect some human activities and work. Umachandran (2020) stated that Robotics technology is an interdisciplinary field of information engineering, computer science and electro mechatronics to compute and control using sensory feedback and dispensation of data to support applications that substitute human actions. McCoy and Yeganeh (2021) also stated that construction robotics entails the design, construction, operation and application of robots at the different levels of components, building and infrastructure. Turner et al. (2021) opined that Robotics in the construction industry is applied using unmannered aerial vehicles and quadruped robots. Also, robotics is used in various ways, such as surveying construction sites, gathering data for 3D building modelling, and aiding workers in carrying heavy tool kits. Robots, through various computer programming, can be used for painting, bricklaying, loading using an autonomous vehicle without an operator, and acts as technical assistant to construction manager and workers. Robots are both use off-site and onsite, where off-site applications include the production of concrete, brickwork or steel components. In contrast, the onsite applications include reinforcement manufacturing and positioning, steel welding, distribution of concrete, façade operations, earthmoving and material handling (McCoy & Yeganah, 2021). The use of Robots in the construction industry brings about: time efficiency, accessibility and decentralisation of information, aids mobility, improves workers' performance, efficient site management, cost-effectiveness, automation of excavation, drilling, earthworks, precision and accuracy, and increment in productivity. However, there are different kinds and types of robots used in the construction industry, which depends on the task for which it is built. Examples are road bricklaying robots, demolition robots, welding robots, painting robots, humanoid robots, exoskeleton suits robots, forklift robots, contour crafting robots, etc. According to Das (2021), typical robots must exhibit five major fields. The five major fields are: operator interface (a medium of communication between the user and the robots); mobility (that is, it must be able to move, and this is usually designed based on human anatomy), manipulators and effectors (that is the ability of the robots to interact with the environment in moving and picking objects according to commands); programming (this the language inputted into the robots to give commands, allowing it to adapt and learn changes in the environment) and Sensors (the ability of the robot to sense and gather data from the environment using sensors). Robotic technology is a construction 4.0 technology that is bringing about change in the construction industry landscape, construction activities, processes and construction management.

1.1.7 Internet of Things (IoT)

Internet of things is an ecosystem of the Internet, where the Internet is the connector between things and things, people and people and also things and people. Umachandran (2020) stated that IoT technology is an integrated computing device through the attachment of sensors to machines, devices and gadgets for data collection and analysis, situation monitoring and recording. The main backbone technology in the IoT system is Radio Frequency Identification (RFID), which consists of a radio transponder, radio receiver and transmitter to automatically identify and track tags attached to objects, gadgets, machines or equipment.

Moreover, Turner et al. (2021) opined that IoT devices use different communication protocols for sending and transmitting data with different frequencies and ranges. These protocols are; Bluetooth with 2.4 GHz frequency and 50–150 nm; Zig Bee 3.0 protocol of 2.4 GHz within 10–150 m; Wifi of 2.4 GHz and 5 GHz band with approximate 50 m; LoRaWAN of different frequency and between 2 and 15 km and also cellular protocols, with a frequency of 900/1800/1900/2100 MHz and with ranges of 85 km (GSM) and 200kkm (HSPA). Furthermore, McCoy and Yeganeh (2021) stated that the IoT technologies, in collaboration with other technologies in construction 4.0, are applied to: monitor and control the progress of construction projects, identify real-time material and equipment location, monitor workers health, knowledge of the material quantity, observe the conditions of equipment on-site and monitoring of vibration, deformation, tensile stress, comprehensive stress, temperature and wind speed in construction sites.

The technologies of construction 4.0 are efficiency optimisers for the Construction industry, which will help shape all aspects of construction activities. Sawhney et al. (2020) stated that adopting Construction 4.0 technologies in the construction industry would solve productivity stagnancy and low efficiency facing the industry. Also, Turner et al. (2021) opined that construction 4.0 technologies adoption in the industry would lead to efficient utilisation of natural resources, reduced life cycle, costs of construction projects and reduction in greenhouse emission. Qureshi et al. (2020) stated that construction 4.0 would solve the construction industry's long-standing challenges of conflict in scheduling, cost/ budgets overruns and construction risk management. Hossain and Nadeem (2019) opined that the digitalisation of the construction industry would introduce: improvements in construction designs, quality enhancements of construction projects, and efficient digitalisation of the construction industry value chain. Osunsanmi et al. (2018) stated that the adoption of Construction 4.0 technologies would have the following advantages: the creation of a sustainable building, speedy budget project delivery, enhancement of safety, facilitation of harmonious relationship between professionals, adequate construction planning, effective monitoring and controlling, encourage seamless communication throughout the organisation, promotes the participation of employees in the decision making process, reduction of errors and rework, and overall ensure client satisfaction. Unequivocally, the adoption and usage of Construction, 4.0 technologies in the construction industry and the built environment will lead to: a sustainable environment, high rate of project success, cost-effective construction projects, minimisation of waste, improvement in safety management, assurance of quality at all stages of construction, enhanced time performance and schedule, aid more innovations breakthrough in the industry, improved workers productivity, increase the life cycle value of buildings and construction projects, effective data management and analytics, helps in the attainment of the United Nation 2030 sustainable development goals and increase construction industry contribution to a nation's economic growth and development.

1.1.8 The Need for Reskilling and Upskilling

However, the benefits, importance, features, and advantages of Construction 4.0 technologies adoption in the construction industry have placed an enormous demand on skill development in the industry since the industry is an epochal transition. Adepoju and Aigbavboa (2021) and Alaloul et al. (2019) stated that these technologies would lead to skill disruptions altering skills, practices and approaches in the construction industry. Also, this disruption will lead to the loss of thousands of jobs, as the technologies of construction 4.0 will replace human labour. According to Adepoju and Aigbavboa (2020), the digitalisation of construction work comes with several benefits and opportunities despite its challenges, such as the loss of jobs. The benefits and opportunities identified are: advancements in productivity of employees, promotion of job satisfaction among workers, reduction in construction risks for workers, reduction in the number of hours required for a worker, reduction in accidents and loss of lives, workers access to current project information and team management.

Nevertheless, the challenges and opportunities ushered by construction 4.0 technologies are a research gap in skill development in upskilling and re-skilling in the construction industry, which is the book's focus. Slowey et al. (2019) and Cerika and Maksumic (2017) also emphasised the opportunity of skill acquisition and development in creating a skilful, better and creative workforce for the industry. Also, Ras et al. (2017) opined that technology usage in industry 4.0 would lead to a shift and upgrade of job profiles, which might be challenging since most needed skills are more interdisciplinary. Manda and Backhouse (2017) stated that the new technologies in the industry require a new breed of innovative and technologically savvy workers.

Vuyiswa and Nischolan (2019) noted that the new required skills would include: analytical thinking and innovation; active learning and learning strategies; creativity, originality and initiative; technology design and programming; critical thinking and analysis; complex problem solving; leadership and social influence; emotional intelligence; reasoning, problem-solving and creativity; and system analysis and evaluation. According to a McKinsey report, more than 371 Million workers might need to change their levels and type of skills by the year 2030 due to the digitalisation of various industries (Chakma & Chaijinda, 2020). The required skills are technical and soft skills for utilising each of the technologies in the industry, which will lead to a wide skill gap in the construction industry. Adepoju & Aigbavboa (2021) averred that the skill gap entails the differences in the current level of skills in an organisation or industry and the future skills needs of the organisation. This is currently happening in the construction industry before the advent of construction 4.0, where employers cannot find the required work with the right skill in the labour market even in developed and developing countries (Juricic, Galic & Marenjak, 2021). However, in bridging this skill gap created by construction 4.0, there is a need for re-skilling and upskilling, which construction stakeholders and professionals must support across the industry.

According to Sivalingam & Manson (2020), re-skilling means learning an entirely new skill which leads to a new job profile and career position while upskilling is learning new skills that will improve the current job profile. Also, as stated in Chakma and Chaijinda (2020), the, Cambridge dictionary describes re-skilling as an attempt made "*to learn new skills so that you can do different jobs*" and upskilling means "*to learn new skills or to teach workers new skills*". Therefore, with construction 4.0 technologies in the construction industry, construction workers and professionals must adapt quickly to the mechanism of upskilling and re-skilling of skills in the industry to match the expected Construction 4.0 technical and soft skills. This is necessary to avert the loss of an experienced and knowledgeable construction workforce.

Consequently, in re-skilling and upskilling of skills in the construction industry in line with construction 4.0 technology, there is need for a collaborative though separated efforts of construction industry stakeholders, including the industry, academia and the government, in the form of a skilled supply chain. Through various construction companies, the industry must begin to see re-skilling and upskilling of their workforce (through training and retraining) as an investment into organisational growth and sustainability. Construction companies must start to give skill training a priority by ensuring their professionals and workers attend required courses, conferences, symposiums, and events that will re-skill and upskill them in line with the skills needed for construction 4.0. Also, construction companies should adopt the mechanism of interdepartmental and Intra–organisational partnership for appropriate knowledge and skill transfer, thus helping them in re-skilling and upskilling their workforce. However, as the industry through construction companies and bodies shapes the current workforce in line with construction 4.0 required skills, academia has a two-way approach to help in re-skilling and upskilling in the construction industry. Firstly, academia must rejig curriculum in the built environment to suit the requirements and skills for construction 4.0. Graduates from the built environment field should naturally fit perfectly into the construction 4.0 landscape. This will further boost re-skilling and upskilling initiatives in the construction industry as work/ industry integration will be easy and done on time in achieving construction companies' goals. Also, academia should focus on more research in adapting the needed Construction 4.0 skills, as the book has done. Research into required skills, which are different according to the types of construction 4.0 technologies, will aid easy re-skilling and upskilling as the assimilation, retention and practicalizing process will be easy, quick and balanced. Vinayan et al. (2020) also stated that a partnership between industry and academic institutions would be cost-effective and make skills retention faster. This will also ensure that researches into various Construction 4.0 technologies and skills needed are based on industrial need, whereby researchers wouldn't be vaguely researching. Furthermore, the government, which will be the biggest beneficiary of re-skilling and upskilling for construction 4.0, needs to implement policies and regulations to support these skill acquisitions and development processes. This is also to spur technological innovations and simplify technological processes in the industry, going by the significant and colossal role the construction industry plays in the nation's socio-economic development.

1.2 Aims and Objectives

This book explores the need for re-skilling and upskill in the construction industry occasioned by construction 4.0 technologies and analyse the roles of construction industry stakeholders such as academics, industry and Government in enhancing Construction 4.0 skills. The book objectives are summarised as follows;

a. To explore the various Construction 4.0 technologies and their applications in the construction industry.
b. To examine the benefits and opportunities provided by different construction 4.0 technologies.
c. To identify the various risks and threats inherent in the Construction 4.0 technologies.
d. To analyse the relevant skills in each of the Construction 4.0 technologies.
e. To develop a framework for re-skilling and upskilling human resources in the construction industry to meet the needed construction industry skills for construction 4.0.

1.3 Organization of the Book

For clarity, simplicity, and understanding, this book is structured into three (3) parts. Part A introduces the of Construction 4.0 technologies giving its conceptual and historical background, the prospects of the construction 4.0 in the industry, and the technologies' threats. Part A has two (2) chapters. Part B has six (6) chapters explaining the different technologies of construction 4.0, delineating the applications, benefits, risks and skills of the various technologies. Part C has only One (1) chapter focusing on re-skilling in the construction industry according to the different technologies. Part C also highlights the role of the major stakeholders in the construction industry on the need for re-skilling concerning Construction 4.0 technologies. The book was concluded in Part C, summarising the significant highlights of the book and the book's essence.

1.4 Contribution and Value

This is a highly innovative and timely book for the construction industry and built environment generally in utilising Construction 4.0 technologies. The contribution and value of this book are in five (5) folds. Firstly, the book bridges the gap of the relevant skills needed in construction 4.0, where many pieces of research have given a general perspective of the skills required. Still, this book explains the skills needed for each of the technologies in construction 4.0. This, of course, will give practical insights into skill development and help strengthen specialisation in the construction industry, where different construction companies can access the needed skills according to their technological strengths and Construction 4.0 adoption. Also, this book highlighted the various risks of construction 4.0 technologies, not on a general basis but on each of the technologies. This will aid skill sustainability and provide the modus operandi for utilisation in the industry, where construction 4.0 will not do more harm than good. Thirdly, the book is a cure to collaborative deficiency among the stakeholders in the industry in developing the needed skills for the epochal transition to construction 4.0. It brings out the various roles of the three major stakeholders in the industry on relevant, necessary and significant skills needed to implement the technologies of construction 4.0. The fourth value this book contributes is the various case studies on construction 4.0 technologies, which gives a practical insight into how the technologies work, which will ensure skill building confidence for construction professionals. Lastly, the book introduces strategies for re-skilling and upskilling in the construction industry to meet the demand for construction 4.0.

1.5 Summary

This chapter gives an extensive introduction to Construction 4.0 and the major technologies. It highlighted the birth of construction 4.0 from industry 4.0 and provides a background into the major technologies of construction 4.0. The chapter also introduces the essence of re-skilling and upskilling in the industry and how the major stakeholders can contribute to skill development in the industry. This chapter also highlighted the aims and objectives of the books and the contributed values of the book.

References

Adepoju, O. O., & Aigbavboa, C. O. (2020). Implementation of construction 4.0 in Nigeria: Evaluating the opportunities and threats on the workforce. *Academic Journal of Interdisciplinary Studies, 9*(5), 254–264. https://doi.org/10.36941/ajis-2020-0102.

Adepoju, O. O., & Aigbavboa, C. O. (2021). Assessing knowledge and skills Gap for construction 4.0 in a developing Economy. *Journal of Public Affairs* 1–10. https://doi.org/10.1002/pa.2264.

African Review. (2021). Africa's construction sector to grow at a CAGR of 6.4% by 2024. https://www.africanreview.com/construction-a-mining/buildings/africa-s-construction-sector-to-grow-at-a-cagr-of-6-4-by-2024.

Alaloul, W. S., Liew, M. S., Zawawi, N. A., & Kennedy, I. B. (2019). Industrial revolution 4.0 in the construction industry: Challenges and opportunities for stakeholders. *Ain Shams Engineering Journal, 11*(1). https://doi.org/10.1016/j.asej.2019.08.010.

Allouzi, R., Al-Azhari, W., & Allouzi, R. (2020). Conventional construction and 3D printing: A comparison study on material cost in Jordan. *Journal of Engineering*, 1–14. https://doi.org/10.1155/2020/1424682.

Auti, S. D., & Patil, J. R. (2018). Prefabrication technology—A promising alternative in construction industry. *International Journal of Science and Research (IJSR), 8*(8), 240–244.

Cerika, A., & Maksumic, S. (2017). The effects of new emerging technologies on human resources: Emergence of industry 4.0, a necessary evil? Faculty of Business and Law, Unversity of Agder, pp. 1–73.

Chakma, S., & Chaijinda, N. (2020). Importance of reskilling and upskilling the workforce. *Interdisciplinary Sripatum Chonburi Journal, 6*(2), 23–31.

Chapman, A. (2013). Drone types: Multi-Rotor vs Fixed-wing vs Single Rotor Vs Hybrid VTOL. AUAV. https://www.auav.com.au/articles/drone-types/. Accessed 25 May 2021.

Cherdo, L. (2021). The 13 best construction 3D printers in 2021. ANIWAA. https://www.aniwaa.com/buyers-guide/3d-printers/house-3d-printer-construction/. Accessed 25 May 2021.

Das, S. (2021). What are the 5 Major fields of Robotics? The coderworld. https://www.thecoderworld.com/what-are-the-5-major-fields-of-robotics/. Accessed 30 Apr 2021.

Formlabs (2021). Powerful, Affordable 3D Printers for Professionals. Formlabs. Available at https://formlabs.com/3d-printers/ (Accessed May 25, 2021).

Hermann, M., Pentek, T. & Otto, B. (2016). Design principles for industry 4.0 scenarios. In: *49th Hawaii International Conference on System Sciences* (pp. 3928–3937).

Hossain, M. D., & Nadeem, A. (2019). Towards digitising the construction industry: State of the art of construction 4.0. *Interdependence between Structural Engineering and Construction Management (ISEC), 6*(2), 1–6.

Hossain, M. A., Zhumabekova, A., Paul, S.C., & Kim, J.R. (2020). A review of 3D printing in construction and its impact on the Labour Market. *Sustainability, 12*(8492). https://doi.org/10.3390/su12208492.

Juricic, B., Galic, M., & Marenjak, S. (2021). Review of the construction labour demand and shortages in the EU. *Buildings 11*(17), 1–17. https://doi.org/10.3390/buildings11010017.

Kozlovska, M., Klosova, D., & Strukova, Z (2021). Impact of industry 4.0 platform on the formation of construction 4.0 concept: A literature review. *Sustainability, 13*, 2683. https://doi.org/10.3390/su13052683.

Li, Z., Shen, G. Q., & Alshawi, M. (2014). Measuring the impact of prefabrication on construction waste reduction: An empirical study in China. *Resources, Conservation and Recycling, 91*, 27–39. https://doi.org/10.1016/j.resconrec.2014.07.013

Manda, M. I., & Backhouse, J. (2017). Digital transformation for inclusive growth in South Africa: challenges and opportunities in the 4 th industrial revolution. 3rd African Conference on Information Systems and Technology, Cape Town, South Africa, July 2017.

Markets & Markets. (2020). 3D concrete printing market to reach $56.4 Million by 2021. https://www.3printr.com/3d-concrete-printing-market-reach-56-4-million-2021-1239664. Accessed 24 May 2021.

Mays, P. (2021). Architecture, engineering & construction prefabrication and industrialised construction. Dassault System.

McCoy, A., & Yeganeh, A. (2021). An overview of emerging construction technologies. NAIOP Research Foundation, pp. 1–40.

Mohandes, S. R., & Omramy, H. (2013). Building information modeling in construction industry: Review paper, Conference proceedings of International Conference of Seminar Kebangsaan Aplikasi Sains & Mathematic (SKASM), Malaysia, October 2013.

Ocean, J. (2020). Top BIM construction software tools in 2021. Revizto. https://revizto.com/en/best-bim-software-tools-2021/. Accessed 22 May 2021.

Osunsanmi, T. O., Aigbavboa, C., & Oke, A. (2018). Construction 4.0: The future of the construction industry in South Africa. *World Academy of Science, Engineering and Technology, International Journal of Civil and Environmental Engineering, 12*(3), 150–156.

Qureshi, A. H., Alaloul, W. S., Manzoor, B., Musarat, M. A., Saad, S., & Ammad, S. (2020). Implications of machine learning integrated technologies for construction progress detection under industry 4.0 (IR 4.0). In: *Second International Sustainability and Resilience Conference: Technology and Innovation in Building Designs*. https://doi.org/10.1109/IEEECONF51154.2020.9319974.

Ras, E., Wild, F., Stahl, C., & Baudet, A. (2017). Bridging the skills gap of workers in industry 4.0 by human performance augmentation tools—Challenges and roadmap. In: *Conference Proceedings of the Tenth International conference on Pervasive Technologies Related to Assistive, Island of Rhodes, Greece, June 21–23* (pp. 428–432). https://doi.org/10.1145/3056540.3076192.

Sawhney, A., Riley, M., Irizarry, J., & Perez, C. T. (2020). A proposed framework for construction 4.0 based on a review of literature. *EPiC Series in Built Environment, 1*, 301–309.

Sculpteo. (2021). Top 16 of the best 3D modeling software for architecture in 2021. 3D Learning Hub. https://www.sculpteo.com/en/3d-learning-hub/3d-printing-software/best-architecture-design-software/. Accessed 20 May 2021.

Shahrubudin, N., Lee, T. C., & Ramlan, R. (2019). An overview on 3D printing technology: Technological, materials, and Applications. In: 2nd International conference on sustainable material processing and manufacturing (SMPM 2019). *Procedia Manufacturing, 35*(2019), 1286–1296.

Sivalingam, A. D., & Mansori, S. (2020). How sorganisations should view re-skilling and upskilling the workforce. https://www.sastraeducation.com/how-organizations-should-view-reskilling-and-upskilling-the-workforce-.html.

Slowey, K., Cowin, L, Beeton, J., & Brown, K. (2019). 7 Trends that will shape commercial construction in 2019, January 7, 2019. https://www.constructiondive.com/news/7-trends-that-will-shape-commercial-construction-in-2019/543978/.

Tahir, M. M., Haron, N. A., Alias, A. H., Al-Juma, A. T., Muhamad, I. B., & Harun, A. N. (2017). Applications of building information model (BIM) in Malaysian construction industry. In: *International Conference on Architecture and Civil Engineering (ICACE), IOP conference series; Material science and Engineering, 291 (2017) 012009* (pp. 1–7). https://doi.org/10.1088/1757-899X/291/1/012009.

Tkac, M., & Mesaros, P. (2019). Utilising drone technology in the civil engineering. SSP—*Journal of Civil Engineering, 14*(1), 27–37. https://doi.org/10.1515/sspjce-2019-0003.

Turner, C. J., Oyekan, J., Stergioulas, L., & Griffin, D. (2021). Utilising industry 4.0 on the construction site: Challenges and opportunities. *IEEE Transactions on Industrial Informatics, 17*(2), 746–756. https://doi.org/10.1109/TII.2020.3002197.

Umachandran, K. (2020). Industry 4.0 and its associated technologies. *Journal of Emerging Technologies (JET), 1*(1), 1–10.

Vinayan, G., Harikirishanan, D., & Ling, S. M. (2020). Upskilling and reskilling the workforce via industry driven technical and vocational education and training: Strategies to initiate Industry/Institution partnership in Malaysia. *Journal of Economic Information, 7*(2), 94–103.

Vuyiswa, L., & Nischolan, P. (2019). Industry 4.0 and its impact on skills development on built environment professionals: A South African perspective. Conference Paper, pp. 538–546.

Wingtra. (2021). Why and how to use drones in Construction and infrastructure. https://wingtra.com/drone-mapping-applications/drones-in-construction-and-infrastructure/. Accessed 15 May 2021.

Chapter 2
Construction 4.0

2.1 Introduction

The present-day is increasing in technological innovations. Technology has transformed practically all sectors, and this has provided new ways of doing things in every aspect of human endeavour. The financial world has fintech; the agriculture world has ag-tech, the academic world has ed-techs, the health sector has health-techs, media has also been largely affected by technology. In fact, it will not be a mistake to label a sector "under-developed" if it has not been affected by technology. Now, we've already had three different industrial revolutions, and we are in the fourth presently. The construction industry is currently experiencing transformation by the fourth industrial revolution and its emerging technologies and this is termed construction 4.0.

In retrospect, the first industrial revolution, which spanned from the sixteenth century to the seventeenth century, specifically from 1760 to around 1830, recorded an exponential growth rate with the mechanical industry, which resulted in the development of a slew of newly developed building materials. Architects and engineers utilised cast, Steel, glass and iron to construct buildings that no one would have imagined its existence concerning its frame, form and purpose (Mokyr, 1998), (Fisher, 2015).

The second revolution between (1870–1914) was distinguished from the first due to its complex creativity based on vital knowledge used to design technologies that fuelled the industry with cheaper and efficient bulk production of railroads, telegraphs, electricity, and steel (Mokyr, 1998). This revolution drove the construction industry in terms of the creativity of architectural design and lightening for vertical spaces (Fisher, 2015); new technologies for prefabrication; and the introduction of designs aided by the computer (Bethany, 2018). This brought about different benefits to the industry. The third industrial revolution occurred between the 1970s and 1980s,

O. Adepoju et al., *Re-skilling Human Resources for Construction 4.0*, Springer Tracts in Civil Engineering, https://doi.org/10.1007/978-3-030-85973-2_2

ushered in the Internet and computer-aided system. This stage, driven by information technology, impacted the architectural design software and introduced construction automation activities such as BIM, Lean processes, Revit, 3D, Enterprise-wide Project management etc.

The rate at which technology is improving with the fourth industrial revolution is faster compared to the previous industrial revolutions the world has experienced. The first industrial revolution was characterised by water and steam as the means of industrial production. The steam engine was used in many factories and industries for production, but it was relatively slow. The elements of the first industrial revolution served well. Still, its limitation of time process made engineers push for more advancement. The constant push for advancement ushered in another industrial revolution after the first, which has electricity as its major characteristics; this worked better than water and steam engine. It aided production and work became easier. Though production became easier than in the first industrial revolution, further attempts to improve, the production process was achieved using information and electronics technology. The current revolution, which is the fourth industrial revolution, now builds on the third revolution by expanding the scope of the information technology of the third revolution (Boyes, 2018).

Newer technologies like the Internet of Things, Sensors, Analytics, Social Media, 3D printing, Artificial Intelligence, Augmented & Virtual Reality, and other technologies have contributed immensely to the current vital transformation being experienced by the construction industry. The manufacturing industry is gradually becoming fully digitalised using automated technologies. The innovation is currently introduced in the construction industry known as construction 4.0. This new development has shown great benefits towards improving the construction process. However, full benefits of digitalized construction processes, automation, and in the context of construction projects is yet to be fully explored.

Prefabricated parts or end-product performance is primarily verified as part of construction projects. Building data from planning to delivery with digital processing continuously, on the other hand, can detect anomalies and thus avoid unnecessary rework. Today, the client's delivery contains digital information about the building configuration. This information is essential as wrong information can have negative consequences on operations, management and performance of buildings project. Digital technologies present a better performance of project with the synergy of information and communication technologies. Also, large amounts of data can be digitally processed and accessible throughout construction projects using information and communication technologies (ICT). This demands product for a reliable and correct information in real-time. Building Information Modelling (BIM) is mainly associated with studies on digital processes in the Construction Industry. Still, there are other digital technologies related to construction 4.0 that can improve operations in construction projects. Internet of Things (IoT), 5G, Artificial intelligence, 5G, and augmented reality are all technologies that have potential benefits. Construction projects are known to have a progressive method of production, and most times, the positive outcome of a project is dependent to a larger extent on the knowledge of each of those involved in the Construction work. However, with the advent of

Construction 4.0, automation will take over most of the construction operations and there will be more reliance on the functionality of the machines and technical knowledge of the operator. For example, in the case of prefabrication, precast of building parts, fully prepared modules, and automation techniques are automated construction techniques. The steps involved in these technologies are analogous to those of the manufacturing sector; products should be categorized and correlated with realisation technologies before production begins (Lessing et al. 2015). Modules produced off-site have the potential to increase construction efficiency (Vianna et al. 2017). 3D printing can be used to make parts of a building on or off-site (i.e., additive manufacturing) and can be modified and augmented for certain purposes. These innovations have brought about automated devices that can function in any unpredictable places thereby making construction areas more convenient.

Furthermore, the appearance of 5G networks is proposed to permit various robotic systems to form a construction process in which robots are designed to replace manual labour in repetitive and dangerous construction procedures such as masonry wall construction (Bruckmann et al., 2016). Combining various methods of pre-configured production can personalise each building construction, and thus the risk of product variations and waste is reduced. For instance, in the area of renovation of building, the adoption of digital processes makes it possible and easy to achieve the desired construction pattern without much redesign. In minimising waste during construction, the introduction of digital technology into construction process help to curtail unnecessary construction work caused by deviations from the projected final product. With the use of digital technology in combination with building with robotics, early detection of errors is made possible thereby creating a better control over building designs (Whyte et al., 2016). Consequently, unnecessary construction work caused by deviations from the intended final product is minimized.

Furthermore, emerging technologies such as Artificial Intelligence avails project managers and professionals the opportunity to get accurate materials required for proposed construction projects and data on materials consumed on already finished projects. This information promotes effective planning, execution, prompt delivery of project and reduction of environmental effects. Before production can begin, automated construction methods must be established.

Most industries have evolved over the last few decades, incorporating process and product creativity into the core of their operations. The construction and engineering sector have lagged in technology opportunities that can help boost production and productivity, resulting in reduced labour productivity (Livotov, 2019). Diverse internal and external challenges are responsible for this situation, including the industry's ongoing fragmentation, the challenge of finding the right talent for your specific job descriptions, insufficient links to contractors and suppliers, and insufficient knowledge transfer from one project to another (Craveiro 2019). Despite the big potential in the industry, expanding productivity and profitability can only be actualised through digital technology, new techniques of Construction, and innovations. Building information modelling (BIM), drones, augmented reality, and three-dimensional (3D) scanning are all mature market tools (Gao, 2019). Firms can benefit from incorporating these innovations by increasing productivity, safety, quality, and improving project management. To unleash this potential, a plan for concerted and concentrated inputs in different areas, including operations, technology, personnel, regulation, and others, need development.

The fourth industrial revolution has brought sensor systems, AI machines, smart materials, and digital technologies to the construction industry. Thus, Construction 4.0 is built on two foundations: use of digital technologies in the construction industry and the industrialisation of the construction processes. Digitisation refers to the field of digital data management via the use of the Internet and software. (Wang, Li & Li, 2015). This technology affects the entire industry, including organisations involved, the environment, and the people. An example of this is the increased movement brought about by laptops, tablets, and smartphones, allowing for and enabling privilege to relevant and up-to-date data on digital buildings at any time and from any location. Indeed, digital technologies are recognised as tools that are increasingly being used in the construction industry to achieve successful outcomes in various tasks and processes Finally, switching to new technologies increases productivity, which industries such as Construction seek the most.

The industrialisation of Construction concentrates on automation of Construction, which comprises a recent collection of technologies and processes that will transform how the construction sector is conceptualised (King, 2018). The adoption of these technologies play a vital role in the execution of productive construction processes in many years to come. Manufacturing processes are included because manufacturers can be providers and developers of new solutions and sources of innovation. In short, the construction industry has focused on technological advancements to seek technological innovations that can increase performance through industrial development. Though the concept of Construction 4.0 has only been around for a short time, it appears to be attracting a lot of attention. As a result, there were several attempts to define it, though the definitions are obscure. Most times they adopt theoretical approaches from its origin which is the fourth industrial revolution.

2.1.1 Historical Background of Construction 4.0

Industry 4.0, which primarily focuses on utilising computer and internet systems, has resulted in significant technological advancements in the twenty-first century (Boyes, 2018). However, this evolution has benefited the construction industry, leading to the coining of "Construction 4.0," which has gained prominence in recent years.

The term construction 4.0 was first mentioned by Roland Berger in 2016. It was founded on the realisation that the construction industry is going digital. This digitisation is built on four concepts which are: automation, digital data, access and connectivity. Construction 4.0 is a relatively new term to the construction industry, but it has gained ground and has been developed a lot in the past five years. Construction 4.0 was derived from the conceptual perspective of a bigger idea known as Industry 4.0. It has been dubbed the "Fourth Industrial Revolution" in recent years because it can influence big industries at the development and production levels, dictating the future of productivity growth in Construction. This word was coined in Germany, where the concept of Industry 4.0 was first announced in 2011 as part of the country's high-tech strategy. Considering that the idea is relatively new, and there is still no agreement on a standard definition. Some authors have emphasised the importance of a dynamic and comprehensive definition, describing the fourth industrial revolution as a.

> shift in the manufacturing logic towards an increasingly decentralised, self-regulating approach of value creation, enabled by concepts and technologies such as cyber-physical systems (CPS), Internet of things (IoT), Internet of Services (IoS), cloud computing, or additive manufacturing and smart factories (Hofmann, 2017).

Technological advancements have resulted in cultural shifts known as "industrial revolutions". Since the dawn of the industrial age, mechanisation spanning from the first industrial revolution brought about intensive use of electrical energy, the second industrial revolution, and widespread globalisation through the third industrial revolution (Lasi et al., 2014). Industry 4.0 enhances relating between information, objects, and people by creating physical and digital construction simulations that allow industries to transform their environment into an intelligent production environment. In light of this, it is vital to understand the rationale behind the adoption of Industry 4.0 ideologies and incorporate it into the construction sector, hence birthing Construction 4.0. To give a proper account of this, Web of Science (WoS/ISI) and Scopus scientific databases conducted a survey to determine when Industry 4.0 and Construction 4.0 were first mentioned in the scientific literature. This is significant because, as previously stated, the Roland Berger report linked these two concepts saying that Construction 4.0 is an example of an Industry 4.0 application. Thus, researches were conducted to know how construction 4.0 is connected to industry 4.0. The first article concerning "Industry 4.0" and "construction 4.0" together dates back to September 2014 (Lasi et al., 2014). An assessment was carried out in the article as to how mechanised 3D-printed mock-ups could meet customers' demands, and a pilot project was established for this purpose. Li and Shi published the first article mentioning Construction 4.0 in August 2015, before Roland Berger (Berger, 2016). The paper containing the result of the research but did not outrightly mention construction. The article explicitly mentioned the automation process for monitoring the construction of dams. As a result, "Industry 4.0" was introduced to the construction industry for the first time in 2014, and "Construction 4.0" was first discussed and described as a notion in 2016.

Existing information on Construction 4.0 is currently insufficient due to a lack of an adequate framework resulting in insufficient data analysis. As a result, it is critical to examine this comparatively novel idea from a broader and more pragmatic perspective, anticipating that Construction 4.0 is more than just traditional buildings with modern technology, but describe an innovative way of understanding Construction in light of creativity and higher productivity. Construction 4.0 is based on the same principles as Industry 4.0, but it focuses on and is relevant to the construction industry. As a result, Construction 4.0 refers to all developments connected to integrating future work techniques in terms of techniques, resources, and customers.

Meanwhile, its first stage of Construction is primarily labour-oriented, with few specialised disciplines as is common today. It requires the use of primitive equipment such as shovels, diggers, and hoes. The second stage is driven by mechanisation and machinery such as excavators, conveyors, and cranes. The second stage, which is the mid-twentieth century, was characterised by many specialised disciplines (e.g., architects, project management). At the same time, some automation construction equipment, but site work remained hugely labour-intensive. However, due to cost and time overruns, the resulting productivities are very low (Barbosa, 2013). The third stage is driven by technological innovations, which has given rise to drones and other automated construction equipment. Technology-driven architecture and design (such as BIM), lean processes and systems, increased digitisation of construction activities, and additional software such as CRM, enterprise-wide project management, and so on was all part of the third stage, which began in the late twentieth century.

In essence, four major changes (four industrial revolutions) have occurred and are growing exponentially in the twenty-first century, particularly in the last 8–10 years. The driving force of technology such as: Big Data (including data such as text, voice, image, and video data), Artificial Intelligence, Virtual & Augmented Reality, 3D printing, Internet of Things/Sensors, and other technologies have begun to play an essential role in this revolution as they became more affordable. In the late 1950s, the Internet, information technology (IT), and the widespread availability of personal computers ushered in a new technological revolution in which mechanical and analogue processes were digitalised, and mass manufacturing gave way to customised products (Yanagawa, 2016). The third wave ushered in a new interaction between architecture and technology, posing a challenge to the manufacturing sector (Kolarevic, 2001). Architects began adopting dispersed 3D computer-aided design software as a representational tool to increase precision and expand the scope of their work (Naboni & Paoletti, 2015).

End-user expectations and attitudes have shifted dramatically as well: the new generation is very tech-oriented, enlightened, globally-connected, and uncompromising, demanding world-class quality at low prices and expecting a very memorable personal experience during the home purchase and possession process (Gardiner, 2016). On the one hand, the construction industry's focus is shifting to customisation en-masse, particularly in urban areas, and mass-production rapidly in rural areas. It is pertinent to note that because of changing climatic conditions around the world, there is a unique emphasis on sustainability and green construction, which was previously a low priority (Fig. 2.1).

Most countries are currently operating at the Construction 3.0 level. Nevertheless, due to the construction industry's challenges, there is now an urgent need to transition to Construction 4.0. This new advent of Construction 4.0 presents its own challenges while also presenting new and innovative solutions. This proves that humans have had a substantial effect on the ecosystem of Construction, including the construction industry (Gildow 2012). The essential elements of the revolution in technology can be divided into three categories: material, energy power, and control technology. These elements assess the procedure and strength capacity of the products built, thereby evaluating the paths of human experience, the form and capacity of human

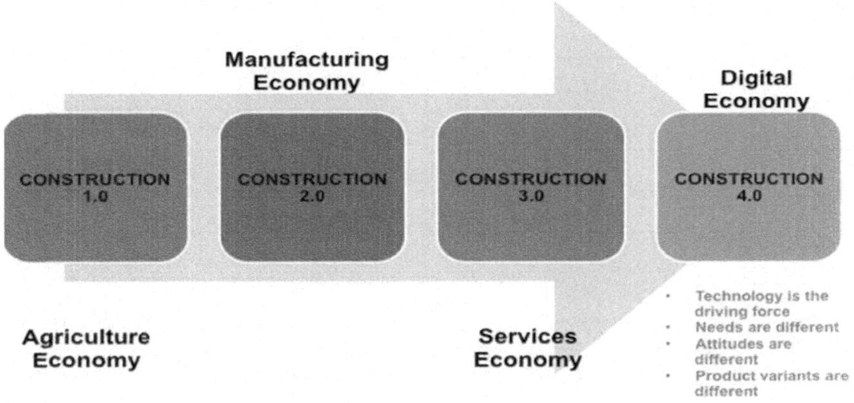

Fig. 2.1 Construction 4.0: The 4th revolution: (Rastogi, 2017)

understanding, and the technique by which humans assimilate information, allowing for the construction of a new living method (Wang, Li & Li, 2015). Previous to the nineteenth century, Construction is limited within the confines of specific properties such as the weight of the building, how high the building is, the strength of the building due to the strength of materials (man-made and natural materials such as: timber, stone, lime mortar, and concrete) used in Construction, (Gildow 2012).

2.1.2 *Components of Construction 4.0*

The component of construction 4.0 in the construction industry can be classified into three (3) sections (Osunsanmi et al., 2018). These include the following:

- The Smart Construction
- Simulation tools and
- Virtualisation construction.

Prefabrication/Modularisation, Internet of Things, Automation, Internet of Services, Product-lifecycle management (PLM), Human–computer interaction (HCI), Addictive manufacturing, Radio-frequency identification (RFID), Robotics, Cyber-physical systems (CPS), and embedded systems are all examples of technologies that enable smart Construction. Building information modelling (BIM) and augmented/virtual/mixed reality are two types of simulation technologies, while virtualised Construction involves the use of mobile computing. Big data, social media, and cloud computing are all buzzwords these days (Fig. 2.2).

(i) **Smart Construction:** Smart Construction seeks to transform the construction industry through the introduction of automation and robotics. This construction approach involves building process under controlled conditions

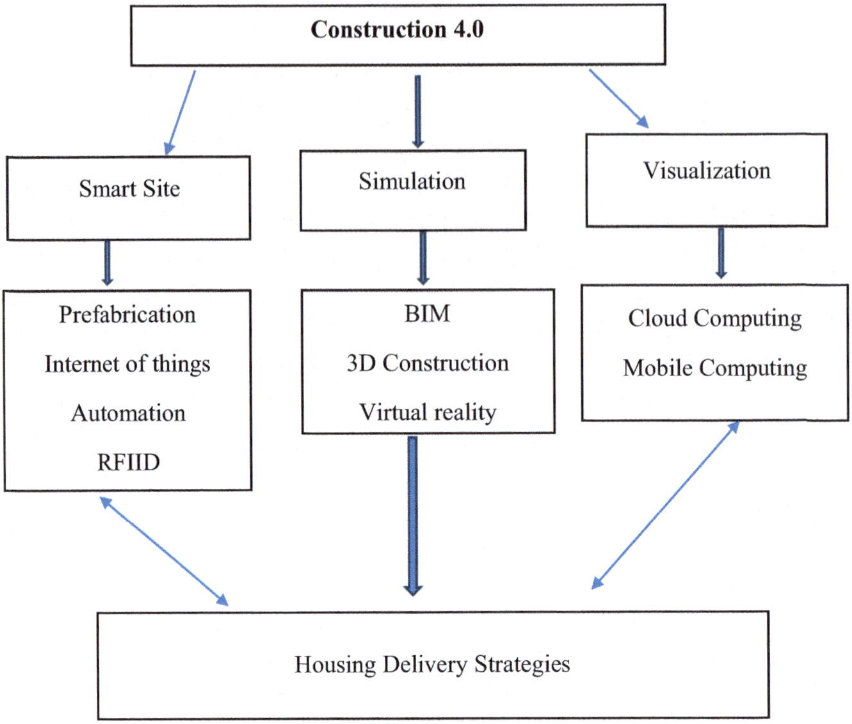

Fig. 2.2 Components of construction 4.0 (Temidayo et al., 2018)

with a high level of resilient which help to minimize waste in construction. This technology gives the sector the opportunity of project costs reduction, waste, increasing precision and sustainability. Smart construction works with intelligent construction objects (SCOs) which are considered smart construction resources. Examples of construction objects include machinery, equipment, materials, and structures with sensor, computation, networking, responding, and communication capabilities which possess. These equipment's have the autonomy and ability to interact with their surroundings to improve decision-making (Niu et al., 2016). Smart Construction involves the use of digital technologies, industrialized techniques and cooperative partnerships to design construction projects. This has availed the construction industry the opportunity of reduced cost, increased productivity and better client's satisfaction.

(ii) **Prefabrication/Modularization:** This is when construction elements and components are manufactured off-site, in a factory-like setting and transported to the job site where it is assembled. Examples of prefabrication are smaller elements in a façade and factory-fitted bathrooms. Prefabrication and modularization are terms that are frequently used interchangeably. Prefabrication is the process of designing construction elements away from the site and assembling them on-site. In theory, these prefabricated elements can be tailored to a specific project. Modular construction refers to instances of prefabrication in which the elements are standardized modules. Each modular unit can be a dorm room, a factory-fitted bathroom, or a façade element (Fig. 2.3).

(iii) **Automation:** Construction automation is a fascinating topic that focuses on using computer-controlled processes and mechanization techniques in the construction industry. It is concerned with applying cutting-edge automated technology to construction projects when human presence is impossible, unwanted, or harmful. The study of construction automation and robots began in the 1980s and has gained prominence in recent years (Son et al., 2010). In a laboratory environment, automation is designed to simulate a building site, produced a winding dry brick semi-autonomously. An automated method of construction is designed with a capacity which permit a self-contained ground robot to create a protective shield utilizing responsive

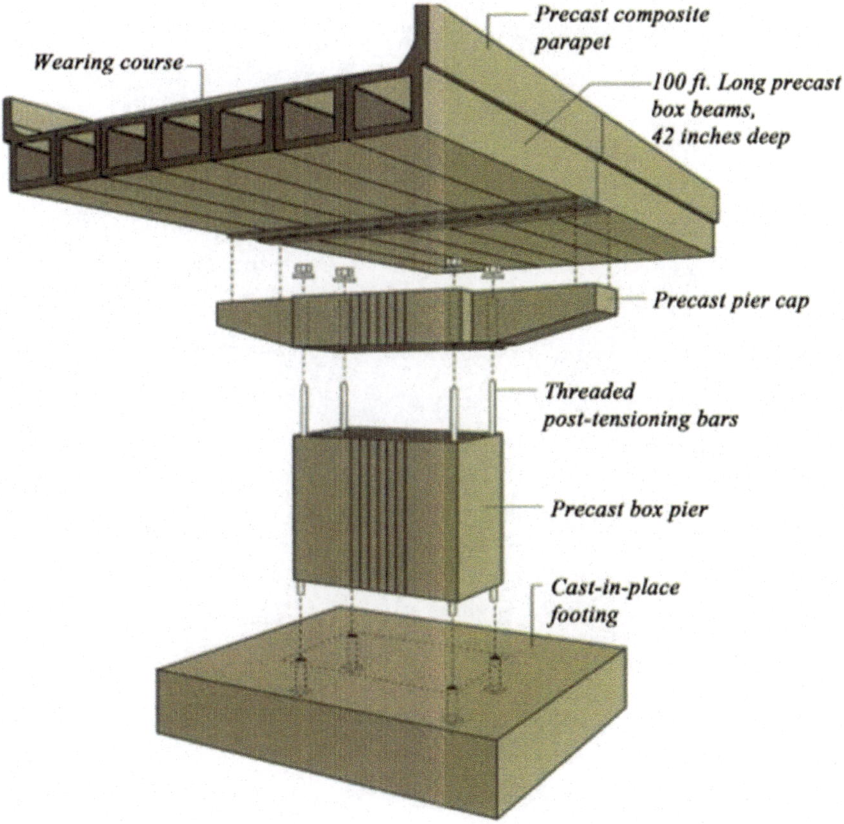

Fig. 2.3 Prefabricated accelerated bridge construction (Mohiuddin, 2015)

pockets (Soleymani et al., 2015). Construction automation and robotics are two examples of cutting-edge technologies in the construction sector. These technologies make use of codes with the use of a combination of mechanical, electronic and computer software to operate robots to carry out specific tasks. These innovations have helped to promote health and safety, better working condition, project scheduling, and improved quality of construction projects. Construction automation uses information technology in conjunction with robotics to aid in designing, preparing, and cost estimation of a project (Hosseini et al., 2018). Furthermore, it uses computers and innovative technologies to resolve challenges in the construction industry, such as shortage of labour. This innovation has rapidly increased the use of prefabricated materials, thereby promoting more of off-site construction methods in the construction industry (Fig. 2.4).

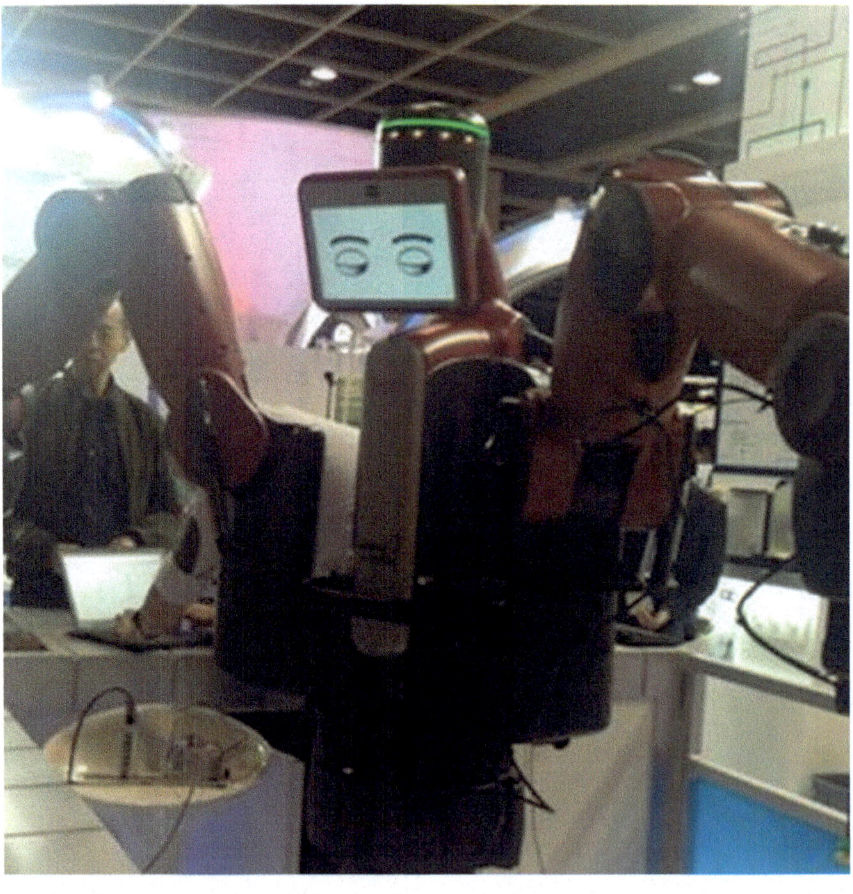

Fig. 2.4 Robotics for the construction industry (Li, 2018)

(iv) **Internet of Things:** The Internet of Things is mainly seen as actual objects that are widely dispersed and have limited storage and processing capacity to enhance the durability, performance, and security of smart cities and their infrastructures (Dalvi et al. 2017). Device connected via the Internet of Things will successfully maintain their actual identities in the online world and receive data from the physical world (e.g., their actual location, the status of their environment), thereby interacting with real-world entities. The Internet of Things (IoT) is an innovative technology that present great opportunity towards achieving smart cities (Fig. 2.5).

(v) **Radio-frequency Identification:** Radio Frequency Identification (RFID) is a form of wireless communication to identify objects that have been tagged. It is used in various industrial applications, from keeping track of items checked out of a library to track items through a value chain. An RFID makes use of a microchip, a powered antenna, and a scanner. Its first commercial products were developed in the 1970s but have only recently become more widely available. It is now more affordable to purchase and adapt due to advancements in communication and energy storage (Paulson 2018). RFID relies on a small electrical appliance, usually a microprocessor, to store information. These sensors are generally tiny, about walnut size, and have a large storage for data. They can generate a processed source of power or capacitors, even if they sometimes don't generate electric power (Oke 2019). The detectors

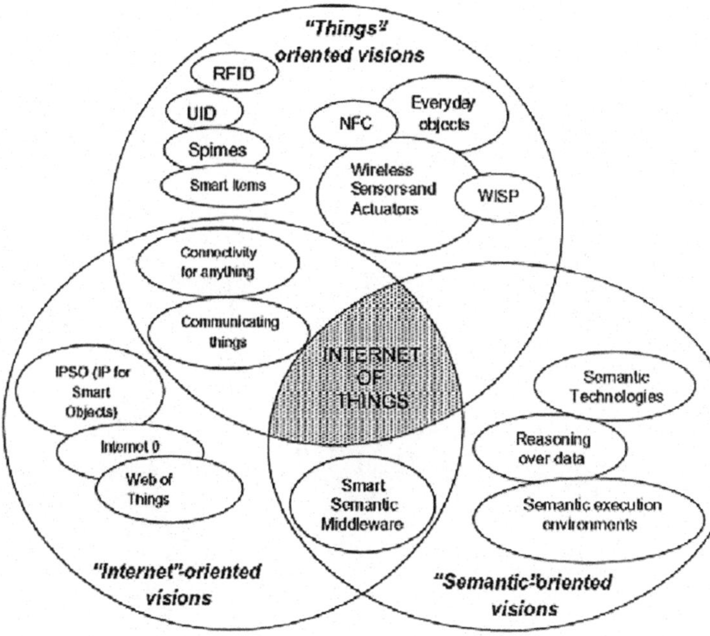

Fig. 2.5 The internet of things: a survey (Atzori et al., 2010)

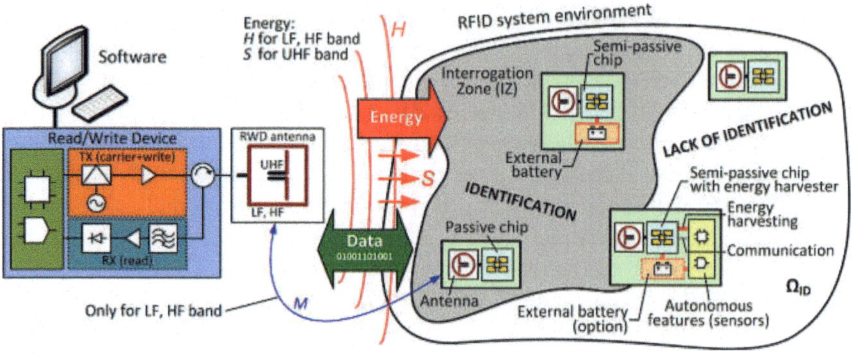

Fig. 2.6 Dynamic RFID identification (Pawłowicz et al. 2020)

which are used to interpret these systems can provide sufficient capacity to read the microprocessor. The technology has various applications, but it is most used to track products, animals, and currency (Fig. 2.6).

(vi) **Robotics:** Construction robots of various types are expected to flood the construction industry in large numbers. For example, there is the 3D-printing robot, which can be used for building large structures on demand. A movable robotic arm is used to run a 3D printer, and the 3D printer uses preset instructions to 3D print a full reinforced concrete building. This technology is also used to build bridges, with the first 3D printed bridge recently completed in the Netherlands. There are machines for laying bricks, cement mixing, tiles laying, wood fitting, and bathroom installations (Fig. 2.7).

(vii) **Simulation:** Construction simulation is a technique used to create and test computer-based images of construction systems in order to better understand the operations mechanism. Professional in the construction industries such

Fig. 2.7 Beginners guide with robotics, (Semra and Hafize 2017)

as architects, designers, builders and engineers adopt construction simulation (4D simulations in BIM) for construction planning and scheduling. This technique avails them the opportunity allows to envision construction arrangements and monitor the life span of a project and progress on the construction site. Furthermore, in construction project planning, simulation offers an accurate project outlook which enables the contractor to analyse all problems encountered during construction with a high level of precision before the commencement of the construction. The emerging technologies with simulation facility help architects and construction designers to assess the advantages of potential solutions without the formation of physically concept. Consequently, prior examination of construction design problems enhanced better decision and reduced cost of project execution.

(viii) **Building Information Modeling (BIM):** Building Information Modelling (BIM) is a process that uses several tools, technologies, and contracts to create and manage virtual models of aesthetic and behavioural aspects of places (Fig. 2.8). (Bello, 2017) Building information models (BIMs) are computer files that can be retrieved, distributed, or networked to aid in the decision-making process for a constructed asset (often but not always in proprietary formats and containing proprietary data). Individuals use the BIM software, businesses, and government organizations to plan, design,

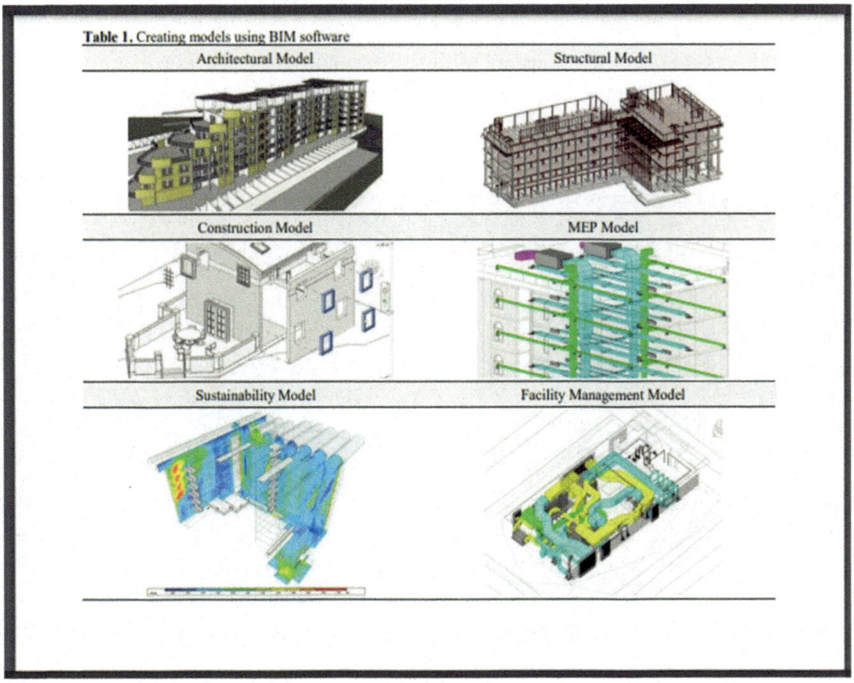

Fig. 2.8 Building information systems and their applications in Turkey (Kalfa 2018)

build, operate, and manage buildings and diverse physical infrastructures such as water, waste, electricity, gas, and communication services, among other things (Ciribini, 2018).

(ix) **Augmented Reality:** Augmented Reality (AR) is a technology that merges the physical and digital worlds, allowing for the interconnection of digital and physical things. (3) The sensory input layered can be either positive that is, beneficial to the natural environment or disruptive that is harmful to the natural environment. Although verifying the building process is complex, detecting real or possible schedule delays in field construction operations is crucial for project administration. The current approach, which does not use AR, is highly demanding, requiring human data collection, interpretation, and graphically complicated representations of progress tracking, among other things (Abdul 2017). As a result, completing this process takes a considerable amount of time, diminishing overall efficiency and increasing resource outflow. Without a doubt, the focus of a real-time feedback and tracking system should be on approaches that can prevent or alleviate the concerns described above. AR is undoubtedly a feasible choice; users may visualize the construction phase in real-time and notice deviations from the original concept (Fig. 2.9).

(x) **Cloud Computing:** Cloud computing is a relatively recent concept in the building sector. It allows for the delivery of a wide range of computer services through the internet and provides construction experts with faster and more flexible resources. The computer services include storage, spreadsheets, networking, servers, application, and analysis. For example, architects and engineers can work on a Building Information Modeling project in an architectural and engineering firm utilizing cloud computing, even

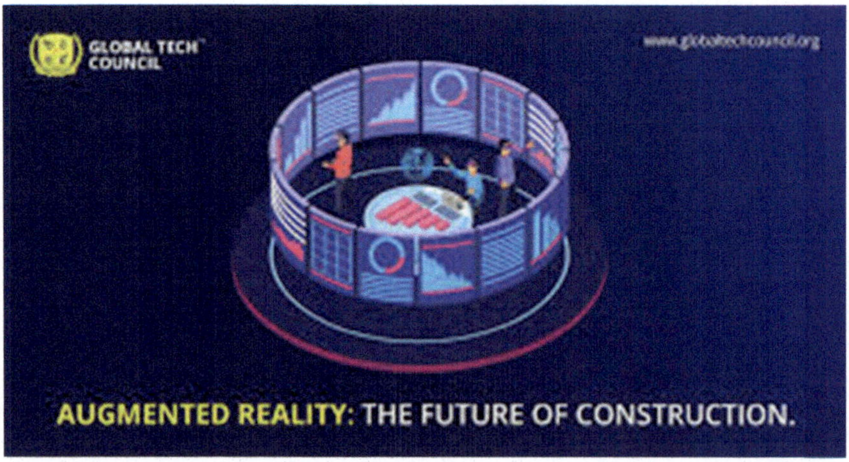

Fig. 2.9 Augmented reality (Toshendra, 2019)

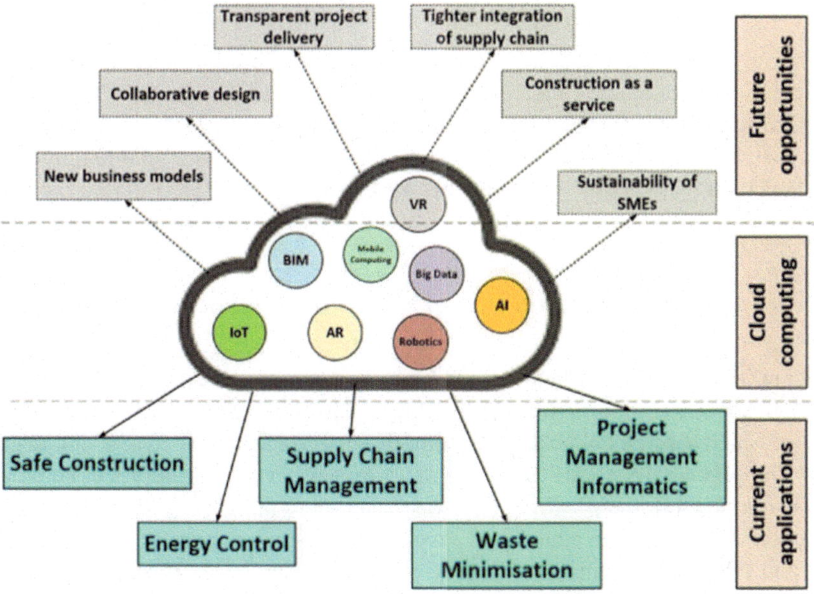

Fig. 2.10 Cloud computing in automation construction (Bello et al., 2021)

if they are in separate places. Since the construction industry is capital-intensive, it can be hesitant to experiment with new technologies, resulting in a slow rate of technology adoption. Cloud computing uses a pay-as-you-go pricing model to include inexpensive and flexible computing resources (Voas, 2009). This cost minimization offer has helped to resolve a major constraint in the adoption of Information Computer Technology in the construction industry. Consequently, cloud computing can be suitable for small projects as it presents a reduced cost of computing facility procurement, installation, and maintenance (Fig. 2.10).

2.2 Prospects of Construction 4.0

With the advent of the Fourth Industrial Revolution, the construction industry can advance in efficiency and productivity as a result of the new technologies. To achieve this, the need to integrate current and new fourth industrial revolution technologies, which are part of the Industry 4.0 model, is vital to ensure a smooth transition into the Construction 4.0 era. The prospects of construction 4.0 are therefore discussed below.

2.2.1 Development of a Creative Environment

Construction 4.0 may have the appropriate mix of accelerators to help the industry embrace an innovative mentality. The invention will lead to integrated solutions that will strike at the heart of the industry's current horizontal, vertical, and longitudinal fragmentation by integrating the real and virtual layers.

2.2.2 Improved Sustainability of the Construction Industry

Building and construction industry takes up 36% of the overall energy and emits nearly 40 per cent carbon dioxide (UNEP 2021). Up till now, the waste from construction sites increases global energy consumption and carbon emissions which negatively impacts on the construction industry. However, with the new technologies of construction 4.0, there will be less waste and more automation in construction. Construction 4.0 is all about using creative technologies to improve the use of materials and use environmental and energy-efficient technologies to achieve sustainability in the construction industry. Construction 4.0 will enable the efficient use of resources for high-quality production while lowering carbon emissions and energy consumption.

2.2.3 Boosting of the Image of the Construction Industry

The building industry's bad reputation is due to several reasons, such as challenging working conditions and lack of automation technology. Construction 4.0's digital and physical technology can enhance the industry's image by transforming the construction work as a whole. This is terms of motivated construction workers, better construction workplaces thereby making the profession more appealing for talent recruitment and retention.

2.2.4 Increased Productivity

Construction is a job that requires focusing on details, and this detailed work makes projects stay long before they are completed. Many times, construction projects take several months before completion despite the large workforce deployed. Construction 4.0 has come in to fill the gap between efficiency and productivity. The technologies employed in construction 4.0 ensure that projects are done with precision within a short time frame, thus increasing productivity and improving efficiency. Meanwhile, the level of productivity matters a lot in construction. It is a combination of having a quality job done within a short period, which construction 4.0 seeks to achieve. Automation can also result in higher quality outcomes. Production will become more consistent and efficient as a result of digitally programming and monitoring manufacturing processes.

2.2.5 Reduction of Risk in Construction Sites

Decisions can be made, and mistakes avoided by creating prototypes and digital models before building begins, thereby reducing financial and reputational risks to clients and contractors. By testing 3D models, designers can evaluate a building's efficiency by examining its energy retention, ability to withstand weather events, and even predicting future maintenance costs.

2.2.6 Enhanced Cost-Predictability

With real-time monitoring, automatic site data collecting, image processing, AI, and analytics technologies, current projects' cost and time prediction can be enhanced. Large volumes of historical data and information can aid in the establishment of standards for initial project time and budget forecast, allowing for longitudinal integrating.

2.2.7 Saves Time

Construction 4.0 technicians create 3D models that allow projects to be organized to improve decision-making processes, avoid inefficiencies and delays, and reduce waste. Construction processes, when done manually, takes time before completion. Sometimes the length of time spent on a construction project matters a lot; some linger to the point that the project is adversely affected. Construction 4.0 automated tools ensure things are done within a short time frame. Automation saves a lot of time by lowering error margins and increasing output speed.

2.2.8 Improves Site Safety

Digitalization has the potential to make construction work safer. Automation frequently reduces physical risks to workers, and the use of augmented and virtual reality eliminates the risks associated with site visits. With improved onsite connectivity, construction professionals can communicate more effectively and avoid unnecessary accidents.

2.2.9 Better Output for Worker's Effort

Construction 4.0 technologies help to make work neater and more durable. Since the workers will be working primarily with automation tools, there is no need to bother about how the building will turn out, as automation makes construction work neat and more beautiful.

2.2.10 Growth in Foreign Trade Communication

Most of the technologies used in the Construction 4.0 will be imported, which means there will be constant communication with foreign manufacturers; this will, in turn, put the country in a good spotlight economically. This importation can also set the tone for the easy import of other goods and seamless export of goods, thus improving the image of the country in foreign-trade business.

2.2.11 Increased Innovation in the Construction Industry

The advent of construction 4.0 will create more automated designs; creating automated designs will lead to more innovation in the industry. Since automation will make construction work faster, there is the tendency that construction workers can have more time to apply their minds to more creative/innovative things.

2.2.12 Increase in National GDP

The construction industry has always been a major contributor to a nation's economy. In the developed country, it accounts for almost 10% of the gross domestic product. The new technologies of construction 4.0 present the benefits of cost reduction, reduction in delay of project execution, increase efficiency and productivity. This is presumed to positively impact on gross domestic product of any nation.

2.3 Threats of Construction 4.0

Construction 4.0 indeed has many prospects for the construction industry; nevertheless, it also poses some threats to the industry. These threats are barriers that create a sceptical stance to construction 4.0 from some that are already used to the old methods of construction.

2.3.1 Reduced Creativity

Since robots and automation will be taking over, the creativity level of construction workers will reduce or be of no use; since materials will come in already prefabricated, they only have to install the prefabricated materials. Thus, the workers do not have to think about designs to use for the construction, and this leads to reduced creativity, which will affect the dynamism in the designs used for construction.

2.3.2 High Start-Up Cost

Some construction companies may find the initial costs of investing in new digital technologies and adopting Construction 4.0 prohibitive. While large corporations may pay for advances in advance, small businesses and self-employed contractors may find it difficult to compete.

2.3.3 Loss of Jobs

Job loss has been cited as a significant impediment to the advancement of construction 4.0. (FIEC, 2019; Fahy, 2019). Particular skills, specific practice, technology preparation, and experience-based judgment are critical in the construction sector. It becomes incredibly tough when this is automated (Alaloul et al., 2018). Automation replaces tasks that should be performed by humans, placing workers at risk of losing their jobs. According to Bayraktar and Ataç (2018), technological developments will alter the labor composition in the short term, resulting in some job losses.

2.3.4 Increase in Social Anxiety Among Different Cadres

The diverse cadres in the construction industry will experience increased social stress due to the demand for new capabilities. Due to their degree of competence, employees with better skills will command higher compensation, whereas individuals with lower abilities will be paid less (Schwab, 2015). Employees who lack the requisite skills to meet the demands of the technology business may become dissatisfied as a result.

2.3.5 Cyber-Security Risk

Cybersecurity is another concern that workers in the fully digitized business will confront. To communicate essential information, the construction sector, like other economic sectors, frequently relies on stakeholders who rely extensively on smartphones and web technologies (Doss et al., 2019). The industry becomes exposed if construction design, architectural plans, and standards are not adequately protected.

2.3.6 Workers Enslavement

Another challenge with construction automation is workers becoming slaves to automated machines and the risk that vast computer data networks will invade the privacy of humans. Consequently, construction workers are at risk if they are not well trained to manage confidential information.

2.3.7 Increase in Unemployment

The introduction of digital technology will increase the unemployment rate in the country as many will lose their jobs and will be left with little or no option as means of survival. An increase in the unemployment rate means a decrease in the country's economic growth.

2.3.8 Increase in Expenditure of Import Materials

Since most of the technology that will be used will be imported, it implies that the country will generally spend more on import items than the former production rate of construction materials; this will be a big blow to the nation's economy.

2.4 Summary

The Industrial Revolution has resulted in several advancements in several aspects of human life. Because of Industry 4.0, the construction industry has taken on a new look. In any economy, the building industry is one of the most active and flexible industrial sectors. This is because it is critical for all nations' social and economic growth as a driver of economic development, providing an environment in which resources such as equipment, materials, labor, and capital are traded to build an economy's infrastructure. Given the industry's relevance to this study, building operations are defined by traditional methodologies and a lack of ICT implementation, resulting in low-quality infrastructure and harming the construction industry's overall performance. Construction 4.0 is based on industry 4.0, the fourth industrial revolution, and thrived in the manufacturing sector. It comprises blending the physical and virtual worlds through the Internet of Things, simulations, and virtualization. According to this study, the necessity to develop a smart construction site and embrace simulation tools and virtualization for construction operations will drive construction 4.0. The study concludes that construction 4.0 has its pros and cons to the development of the construction industry; meanwhile, the advantages out-weighs the disadvantages. Thus, it is a development that should be embraced.

References

Alaloul, W. S., Liew, M.S., Wan Abdullah Zawawi, N. A., & Mohammed, B. S. (2018). Industry Revolution IR 4.0: Future challenges in construction industry opportunities and challenges in construction industry. *MATEC Web of Conferences, 203*, 02010 https://doi.org/10.1051/mateconf/201820302010.

Atzori, L., Lera, A., & Morabito, G. (2010). The internet of things: a survey. *Computer Network, 54*(15), 2787–2805.

Barbosa, G. F., Carvalho, J. (2013). Design for manufacturing and assembly methodology applied to aircrafts design and manufacturing. In *11th IFAC Workshop on Intelligent Manufacturing Systems, The International Federation of Automatic Control*, Sao Paulo, Brazil.

Bayraktar, O., & Ataç, C. (2018). The effects of industry 4.0 on human resources management. In O. Bayraktar, C. Ataç, E. Yıldırım, & H. Çeştepe (Eds.), *Globalization, institutions and socio-economic performance macro and micro perspectives* (pp. 337–360). Peter Lang GmbH. https://www.researchgate.net/publication/329706763.

Bello, S. A., Gazali, W., & Aderounmu, G. A. (2017). Efficient virtual machine placement algorithm. *International Journal of Computer Science, 14*, 68–75. https://doi.org/10.20943/01201705.6875.

Bello, S., Oyedele, L., Akinade, O., Bilal, M., Davila Delgado, M., Akanbi, L., Ajayi, A., & Owolabi, H. (2021). Cloud computing in construction industry: Use cases, benefits and challenges. *Automation in Construction, 122*. https://doi.org/10.1016/j.autcon.2020.103441.

Bethany (2018). How CAD Has Evolved Since 1982. https://www.scan2cad.com/cad/cadevolved-since-1982/#history-of-cad (Accessed on 30 Dec 2018).

Berger, R. (2016). *Roland Berger Digitization in the Construction Industry: Building Europe's Road to "Construction 4.0"*; Roland Berger GMBH: Munich, Germany.

Boyes, H., Hallaq, B., Cunningham, J., & Watson, T. (2018). The industrial internet of things (IIoT): an analysis framework. *Computers in Industry, 101*, 1–12.

Bruckmann, T., Mattern, H., Spengler, C., Reichert, A., Malkwitz, A & König, M. (2016). Automated construction of masonry buildings using cable-driven parallel robots. In *33rd International Symposium on Automation and Robotics in Construction (ISARC 2016)*, pp. 1–8.

Ciribini, A., DeGiuda, G., & Valaguzza, S. (2018). UK procurement framework informs italy's move to mandate bim. *BIM Plus*. http://www.bimplus.co.uk/people/uk-procurement-framework-informs-italys-move-manda/.

Dalvi, S. A., & Shaikh, M. Z. (2017). Internet of things for smart cities. *Imperial Journal of Interdisciplinary Research, 3*(3), 947–951.

Dalvi et al., (2017). Industry 4.0: The Industrial Internet of Things; Apress: New York, NY, USA, 2017.

Doss, A., Arstein, S. E., & Lehr, LLP (2019). Cybersecurity in the construction industry: protecting against a growing threat. https://www.jdsupra.com/legalnews/cybersecurity-in-the-construction-22150/files/74/.html. (Accessed 25 Sep 2019)

Fahy, D. (2019). Industry 4.0: disaster or new beginning for construction workers? https://www.pbctoday.co.uk/news/bim-news/new-beginning-for-construction-workers/53349/ (Accessed 6 Oct 2019).

Federation Industry European Construction (2019). Construction 4.0. http://www.fiec.eu/en/themes72/construction-40.aspx (Accessed 9 Sep 2019)

Fisher, T. (2015). *Welcome to the Third Industrial Revolution: The Mass-Customisation of Architecture, Practice and Education*. Architectural Design: John Wiley & Sons Ltd.

Gardiner, J. (2016). Exploring prefabrication-international study of prefabrication within aerospace, shipbuilding, automotive and construction industries. http://www.architects.nsw.gov.au/download/BHTS/Gardiner_James_exploring_prefabrication.pdf. (Accessed 24 Oct 2016).

Gao, H., Koch, C., & Wu, Y. (2019). Building information modelling based building energy modelling: a review. *Applied Energy, 238*, 320–343, doi:https://doi.org/10.1016/j.apenergy.2019.01.032.

Gildow, C. (2012). M9—Architecture and the industrial revolution on Module 9. *Architecture* https://learn.canvas.net/courses/24/pages/m9-architecture-and-the-industrialrevolution?module_item_id=44477. (Accessed 30 Dec 2018).

Hofmann, E., Rüsch, M. (2017). Industry 4.0 and the current status as well as future prospects on logistics. *Computers in Industry, 89*, 23–34.

Hosseini, M.R., Banihashemi, S., Zaeri, F & Adibfar, A (2018). Advanced ICT Methodologies (AIM) in the construction industry. In D.B.A. Mehdi Khosrow-Pour (ed.), *Encyclopedia of information science and technology, Fourth Edition*, (pp. 539–550). IGI Global.

King, M. (2018). How industry 4.0 and BIM are shaping the future of the construction environment. *GIM International The Worldwide Magazine for GEOMATICS, 31*, 24–25.

Kolarevic, B. (2001). Designing and manufacturing architecture in the digital age. In *Architectural Information Management, Proceedings of the 19th eCAADe Conference* (pp. 117–123). Helsinki University of Technology: Helsinki, Finland

Kalfa, S. M., (2018). Building Information Modeling (BIM) systems and their applications in Turkey, 1:(1), Journal of Construction Engineering, Management & Innovation, pp 55-66, http://doi.org/10.31462/jcemi.2018.01055066

Lee, G., Sacks, R., & Eastman, C. M. (2006). Specifying parametric building object behaviour (BOB) for a building information modelling system. *Automation in Construction, 15*, 758–776.

Li, C. Z., Hong, J., Xue, F., Shen, G. Q., Xu, X., & Luo, L. (2016). SWOT analysis and internet of things-enabled platform for prefabrication housing production in Hong Kong. *Habitat International, 57*, 74–87.

Li, Q., Shi, J. (2015). Dam construction 4.0. *J. Hydroelectr. Eng., 34*, 1–6.

Li, R. Y. (2018). *Robots for the Construction Industry*. https://doi.org/10.1007/978-981-10-5771-7_2.

Livotov, P., Sekaran, A. P., Law, R., Reay, D., Sarsenova, A., Sayyareh, S. (2019). *Eco-Innovation in process engineering: contradictions, inventive principles and methods* (Vol. 9). The Netherlands, Amsterdam: Elsevier Ltd.

Mohiuddin, A. (2015). *Accelerated bridge construction*. Butterworth-Heinemann.
Mokyr, J. (1998). The Second Industrial Revolution, 1870–1914. In C. Valerio (Ed.), *Storia dell'economia Mondiale*.
Naboni, R., & Paoletti, I. (2015). *Advanced Customization in Architectural Design and Construction*. Springer.
Niu, Y., Lu, W., Chen, K., Huang, G.Q & Anumba, C. (2016). Smart construction objects. *Journal of Computing in Civil Engineering, 30*(4), 04015070
Osunsanmi, T.O., Aigbavboa, C., & Oke, A., (2018). Construction 4.0: the future of the construction industry in South Africa. *International Journal of Civil and Environmental Engineering, 12*(3), pp. 150–156.
Pawłowicz, B., Trybus, B., Salach, M., & Jankowski-Mihułowicz, P., (2020) Dynamic RFID Identification in Urban Traffic Management Systems. *Sensors. 2020; 20*(15):4225. https://doi.org/10.3390/s20154225
Rastogi, S. (2017). Construction 4.0: the 4th generation revolution. In *Proceedings of Indian Lean Construction Conference – ILCC 2017*. Institute for Lean Construction Excellence (ILCE): Chennai
Sanchez, et al. (2006). 2014 Describing the changes in architectural information technology to understand design complexity and free-form architectural expression. *J. Inf. Technol. Constr. Itcon, 11*, 395–408.
Schwab, K. (2015). The Fourth industrial revolution. https://www.weforum.org/about/the-fourth industrial-revolution-by-klaus-schwab/files/75/.html. (Accessed 14 Oct 2015)
Semra, A.S., & Hafize, B., (2017). Digital Proceeding of ICOCEE-CAPPADOCIA2017 New Construction Methods and Hybrid Tectonics: Robotics in Architecture, Digital Proceedings of ICOCEE–CAPPADOCIA2017, Turkey, May 8–10, 2017
Soleymani, T., Trianni, V., Bonani, M., Mondada, F., & Dorigo, M. (2015). Bio-inspired construction with mobile robots and compliant pockets, Robotics and Autonomous Systems, 74, Part B, 340–350. *Retrieved from:* https://doi.org/10.1016/j.robot.2015.07.018
Son, H., Kim, C., Kim, H., Han, S. H., & Kim, M. K. (2010). Trend analyses of research and development on automation and robotics technology in the construction industry. *KSCE Journal of Civil Engineering, 14*(2), 131–139. https://doi.org/10.1007/s12205-010-0131-7
Temidayo, O., Aigbavboa,C., Oke, A& Ohiomaha, I. (2018). Construction 4.0: Towards Delivering of Sustainable Houses in South Africa, Proceedings of the Creative Construction Conference, Ljubljana, Slovenia
Toshendra, S. (2019, April 30). *Augmented Reality: The Future of Construction*. Retrieved from Global Tech Council: https://www.globaltechcouncil.org/augmented-reality/augmented-reality-the-future-of-construction/
Voas, J. Z. (2009). Cloud computing: new wine or just a new bottle? Information Technology (IT) Professional. 11 (2) 15–17, https://doi.org/10.1109/
Wang, Y., Li, X., Li, X. (2015). Eco-Topia: Design 4.0 and the Construction of E-Image City. In *Proceedings of the 1st International Conference on Arts, Design and Contemporary Education, Moscow, Russia, 23*, (pp. 414–417).
Whyte, J., Stasis, A., & Lindkvist, C. (2016). Managing change in the delivery of complex projects: Configuration management, asset information and 'big data.' *International Journal of Project Management, 34*(2), 339–351.
Yanagawa, K. (2016). Reindustrializing architecture. *International Journal of Architectural Computing, 14*(2), 158–166.

Part II
Construction 4.0 Technologies

Chapter 3
Building Information Modelling

3.1 Introduction

Building Information Modelling (BIM) is one of the most promising developments in the Architecture, Engineering, and Construction (AEC) industries. It is a process for creating and managing information about a project all through the project lifecycle. The digital description of every aspect of the physical project is considered the resulting output known as a Building Information Model. The Building Information Model creates a centralized data model where all the project stakeholders can work together effectively. BIM is an intelligent model, which provides insights that enhance the planning, designing, construction, and management of infrastructure. The model is incorporated into all stages of construction and includes data collection, construction design, equipment, construction process, renovation, and demolition. The Building Information Model is a data-rich, object-oriented, intelligent, and parametric digital representation of the facility. View and data appropriate to various users' needs can be extracted and analyzed from it. This enables the generation of information that can be used to make decisions and to improve the process of delivering the facility (AGC, 2005). BIM can be used to demonstrate the entire building life cycle, which is characterized by geometry, spatial relationships, geographical details, building item quantities and properties, cost estimates, inventories of materials, and project schedule (Bazjanac, 2006). BIM makes it possible for projects to be constructed virtually before they are physically built. This helps to eliminate several problems and inadequacies that may arise during the construction process. BIM is the main digital planning method in construction, which helps develop a strategy that ensures collaborative working among participants involved in project execution. In essence, since a construction project often requires collaboration between geographically located parties in diverse locations, BIM creates a platform for different participants to exchange and use information (Hossan & Yeoh, 2018). Consequently, it is a highly collaborative process that allows multiple stakeholders and professionals to collaborate on the planning, designing, and constructing

O. Adepoju et al., *Re-skilling Human Resources for Construction 4.0*, Springer Tracts in Civil Engineering, https://doi.org/10.1007/978-3-030-85973-2_3

of a building project. According to Sai Evuri and Amiri-Arshad (2015), BIM has four characteristics based on several views:

(i) **Information Sharing**: Making information available for all participants involved in a project.
(ii) **Information Management**: These are processes that support acquiring, organizing, storing, sharing, and using information for project success.
(iii) **Part of Visual, Design, and Construction**: It is used as an information system, which can help support the processes defined in visual, design & construction (VDC).
(iv) **Extra Dimension to 3D Models**: It provides all the necessary information required by the entire participants involved in a project.

3.1.1 Dimensions of Building Information Modelling

Several dimensions of BIM have evolved and generated additional information, functions, and application of the Building Information Model System. These evolving dimensions bring about improvement in data processing through the incorporation of all information required for a better understanding of a project. According to Worden (2016), Mesaros et al. (2019), there are seven dimensions of BIM. These include:

(i) **3D Model (geometry)**: This is a well-known dimension because it represents the geographical elements of a building structure and the visible part of the BIM model. It is essentially about the creation of graphical and non – graphical information. The component of the 3D model enables stakeholders to visualize the entire structure of a building before the commencement of the project, thereby facilitating prompt risk detection and risk communication. This model has a central data environment where the same information is shared among the stakeholders, enhancing collaboration. For visualization, programs such as Tekla, Revit, ArchiCAD and Vector works are examples of the 3D BIM dimension (Fig. 3.1).
(ii) **4D Model (time, schedule & duration)**: 4D are models that can be described with the connection between the 3D representation of a proposed project with a time schedule (Fig. 3.2). These models represent four-dimensional modeling of information used for planning construction sites with the inclusion of time elements. It presents data on schedule, which estimates the time required to execute each stage of the project life cycle and its completion. 4D offers the benefit of better information sharing, thereby averting costly delays and supporting early identification of risks. The model presents program and visualization information that shows how a project develops successively. The 4D model ensures optimization of plans, construction activities, and team coordination. Examples of 4D scheduling programs include Primavera, Microsoft Project, and Vico Office Schedule Planner.

Fig. 3.1 3D Building information modeling (Rightserve, 2019)

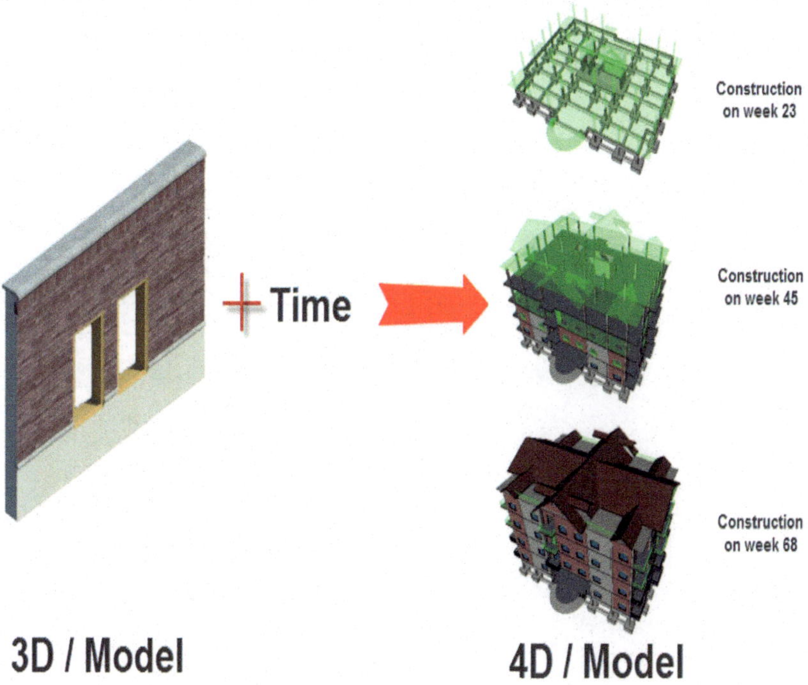

Fig. 3.2 4D model depicting 3D project and schedule (Hijazi et al., 2009)

(iii) **5D Model (Cost estimation and budget analysis)**: 5D is an advancement in 4D with cost-related data. Specifically, this model is used for planning, cost analysis, and budget monitoring. It specifies the costs and the profit in the building budget. With the 5D model, contractors and clients can get updated and accurate cost estimates for projects with adjustments when there is a change in materials, labor, equipment, and scope. A significant advantage of this model is the prompt update of a project's financial status through the 5D accessibility of cost and notification when changes are made. Hence, this simplifies cost and budgetary analysis with predicted and actual expenses over time. It also minimizes budgetary offshoots due to periodic cost reporting, cost monitoring, and cost tracking. Examples of programs for the 5 Costs include Quantity Takeoff and Vico Cost Planner. Figure 3.3. further depict a 5D-BIM application for a project conceptual cost estimating is proposed.

(iv) **6D Model (Sustainability).** The six-dimensional building information modeling is designed to review the energy consumption requirements, which provide the cost implication of the project for sustainability and cost-efficiency. 6D promotes optimization through easy access to the structural details, equipment model, energy requirements, installation date, maintenance plan and configuration technical specification, and estimated lifespan. Majorly, this model provides better insight into the entire project lifecycle,

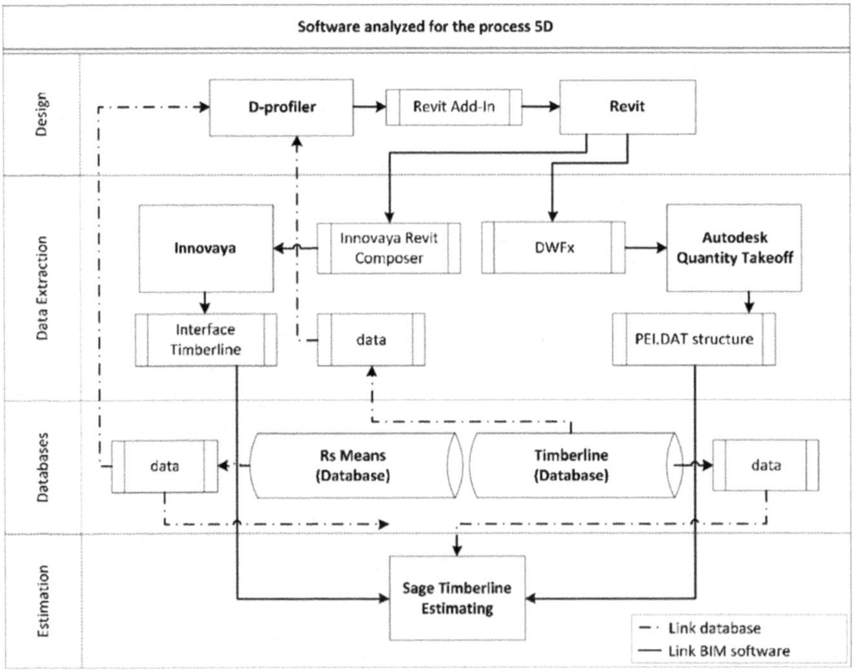

Fig. 3.3 5D Building information modeling cost estimation process (Forgues et al., 2012)

cost, and assets, which enhances decision-making for cost-effectiveness and sustainability. The 6D model makes use of programs such as Eco tech, Green Build Studio, Eco Designer, Hevacomp, ArchiCAD (Fig. 3.4).

(v) **7D BIM (Facility Management)**: The 7D model is used explicitly for facility management, which entails space planning, renovation, and maintenance operations (Azhar, 2015). Project managers utilize this data model for building operations and maintenance. The 7D model is a method that organizes the detailed information of a facility required to manage a building from design to demolition. It presents the benefit of simplifying the repairs, replacement, and maintenance process for contractors and subcontractors (Fig. 3.5).

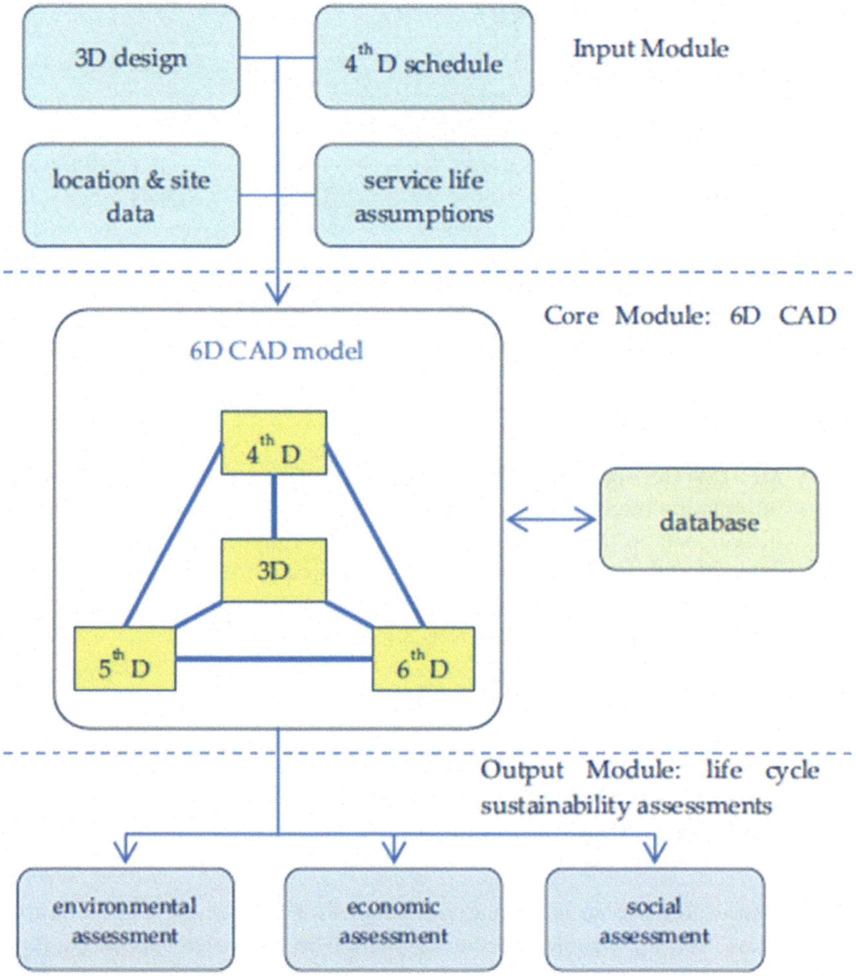

Fig. 3.4 6D CAD model (Yung & Wang, 2014)

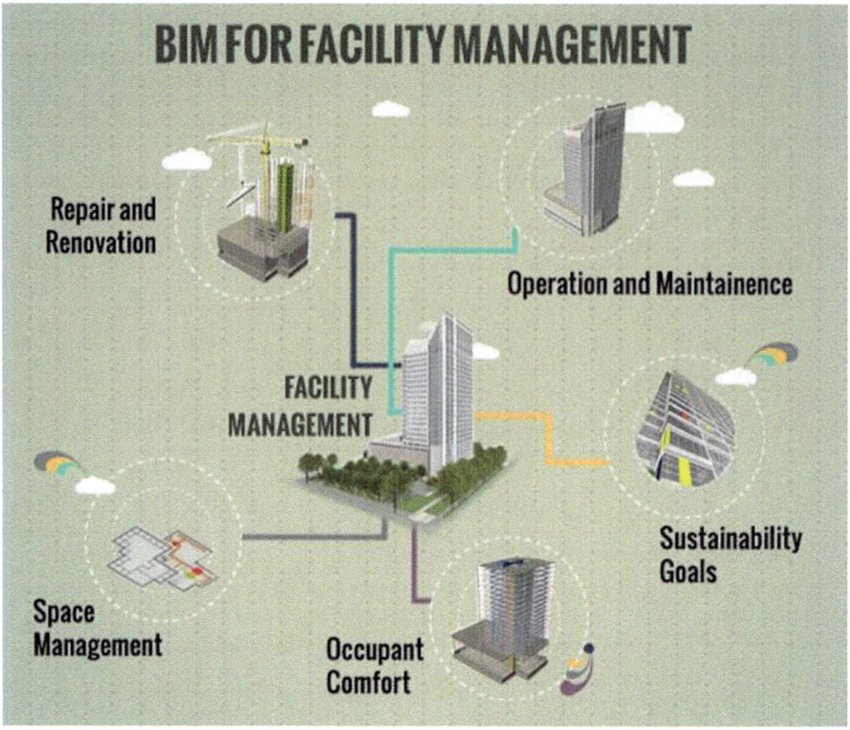

Fig. 3.5 BIM for facility management approach (Weebly, 2020)

(vi) **8D BIM (Occupational safety and hazards)**: The eight-dimensional building information modeling was designed to reduce the incident rate and hazard of accidents in the construction industry. This model is built explicitly for prevention through a design methodology applied to the design process to identify and mitigate risks encountered by construction workers. This model is significant to ensure a safe work environment (Fig. 3.6).

.

3.2 Application of Building Information Models in Construction

BIM has proved to be an intelligent tool adopted to manage the entire construction process. Below are some of the areas where BIM have been applied in project execution (citation).

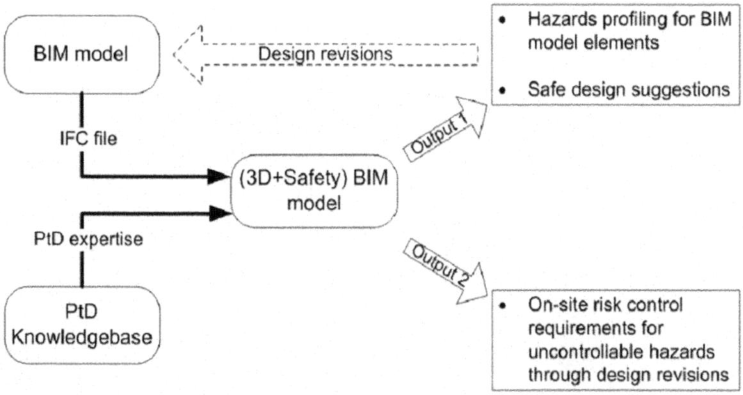

Fig. 3.6 8D building information modeling (Kamardeen, 2010)

3.2.1 *Design Visualization*

BIM has transformed the architectural world by the creation of better and beautiful building designs. One of the crucial roles of the Architects and engineers in the design process is predicting the end result of a building (Boukara & Naamane, 2015). As a result, Architects and designers use visualizations to communicate their concepts with their clients and subcontractors. The use of BIM presents an innovative way of design visualization such as virtual reality, material representation, walk-through, and detection of design errors. All the entire project elements are captured through these data representations, thereby giving the stakeholders an in-depth understanding of the final project before it is commenced. Figure 3.7 shows a BIM visualization of a project with structural components.

3.2.2 *Constructability Analysis*

One of the key functions of Building information modeling is to improve constructability investigation of construction operations. Constructability is a fundamental part of the construction process, and it has been recognized as a critical factor, which may affect the successful execution of a project. It can generally be described as a method of project management used to assess the construction processes from the beginning to the end. Constructability issues often originate due to the gap between the designers' drawings and the contractors' executed projects (Hijazi et al., 2009). Most often, engineers encounter problems with the use of 2D CAD drawings when executing the project. These problems are due to the design of the construction projects and include architectural, mechanical, electrical, and plumbing (MEP) challenges. In situations like this, the project will have to be redone, and this will increase the cost of construction (Yang et al., 2013). BIM-based Constructability investigation

Fig. 3.7 BIM visualization of a structural project (Foster, 2008)

helps evaluate construction processes during the design phase. It identifies possible problems and design errors that might lead to rework delays and an increased cost. Furthermore, constructability analysis is made easier with BIM as it defines the spaces and establishes a suitable relationship between the components of a building model (Hijazi et al., 2009). In summary, the utilization of BIM technology has made it easier to solve the major constructability problems, including design, measurement, clash detection, and space review (Fig. 3.8).

3.2.3 Site Planning and Utilization

Site planning is essential in part of the construction process, affecting project cost, time, and safety on the construction site. The site planning process ensures that all

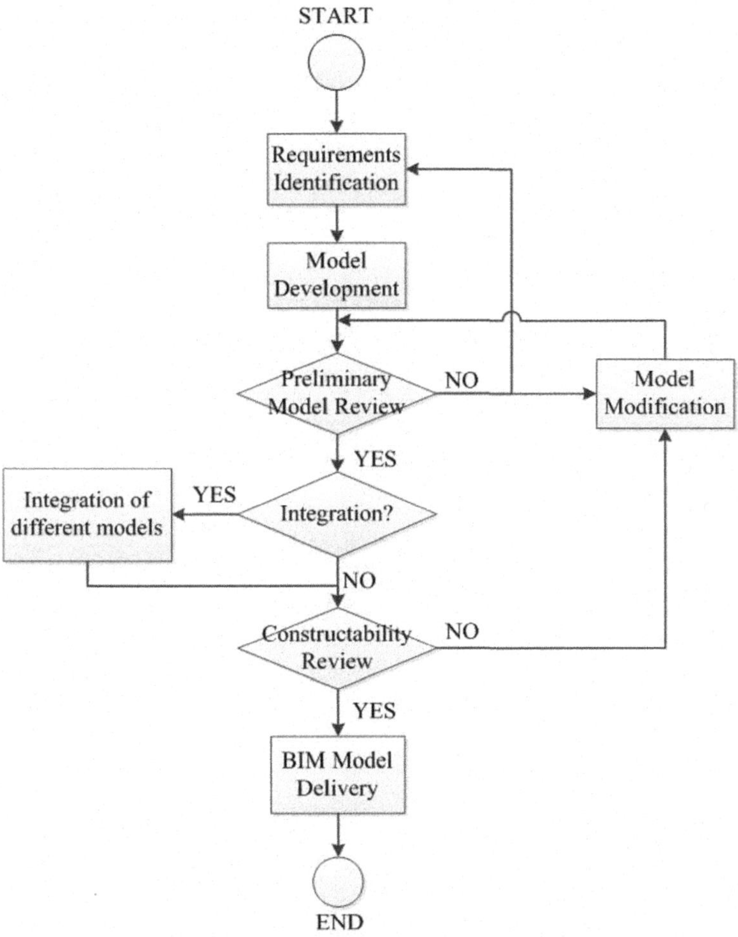

Fig. 3.8 Procedure for constructability using BIM (Yang et al., 2013)

operations and organizations required for the project to progress through the stages of construction are put in place. This includes creating office facilities, worker's chambers, electrification, and storage of materials, transportation, and safety. The BIM-based model offers new opportunities for site planning which provide site layout plans, site safety, and communication. According to Sulankivi et al. (2009), the important contents of a BIM-based site plan include the following:

(i) temporary construction site and surroundings (such as roads, adjoining streets, and parking lots);
(ii) temporary site facilities, structures, and equipment (such as storage facilities, office facilities, fencing, concrete mixer, wood saw, and tower crane);

(iii) temporary site situations, such as area reservations for material storage (such as trash skips with 3D-text, thermal insulators, heating ventilation, and air condition pipes and reinforcing bars); and

(iv) visualizations that promote safety (such as site walkways, crane reach, and vehicles)

Also, construction projects require a lot of temporary facilities to support various construction processes. BIM makes it possible to develop a construction site layout plan that can automate the computation of required materials, assign materials to different storage, and determine the best arrangement of temporary facilities (Kumar & Cheng, 2015). Figure 3.9 depicts a BIM model for construction site layout used for size estimation, creating dynamic layout models and actual travel paths for facility layouts.

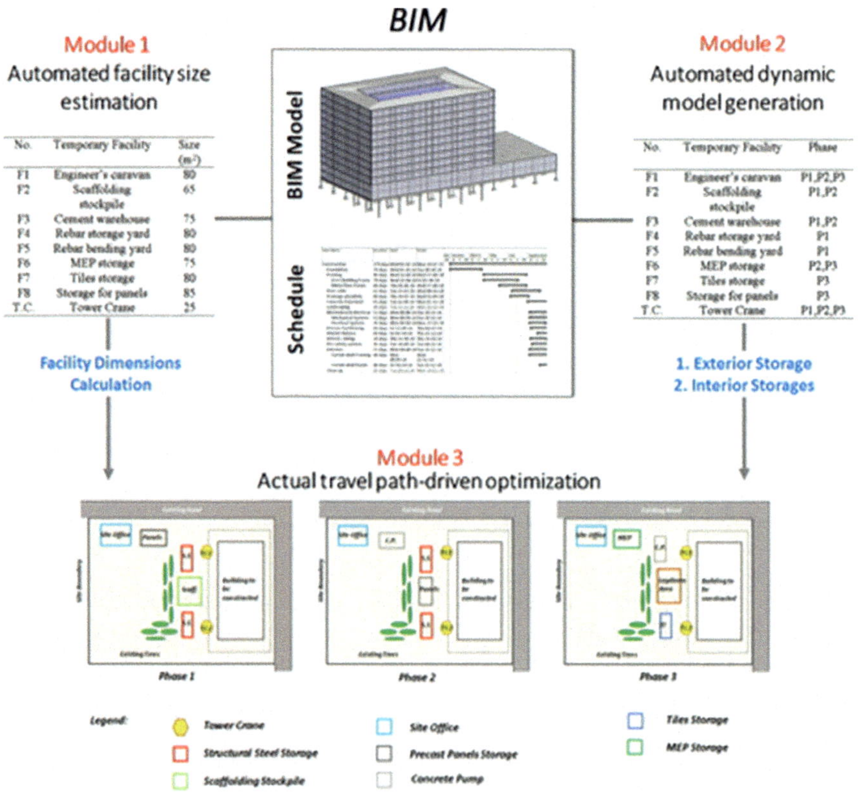

Fig. 3.9 BIM-based automated site layout planning framework (Kumar & Cheng 2015)

3.2.4 Scheduling and Sequencing

Scheduling and sequencing are essential activities that ensure the efficiency of project delivery. These mechanisms are put in place to allocate, communicate tasks, identify relationships among project activities and define the sequence of the tasks for the prompt delivery of a project. BIM provides intelligent networking of a 3D digital model with a built building from start to finish. It assists project participants in confirming if the project is on track, gives contractors a clearer view of the timing and scope of their tasks, and communicates the project's completion dates to the client. In scheduling, it informs team members and tracks the project's progress from the beginning to the end (Barati et al., 2013). BIM also creates a building construction sequence by linking the construction tasks across various professionals, including architects, mechanical, electrical, and structural engineers. In construction sequencing, it helps coordinate material orders, fabrication, and delivery time for all building components (Azhar, 2015). An example is the 4D building information model that comes with an additional time factor function which integrates data with a date for commencing and finishing the supply and installation of construction elements required for the overall project (BIM Community, 2019). BIM ensures accurate project sequencing, which helps identify potential problems and give opportunities to make necessary adjustments without affecting the work of the contractors (Fig. 3.10).

Fig. 3.10 4D construction sequencing (BIM Community, 2019)

3.2.5 Cost Estimation

Accurate cost estimate and effective cost control assessment are essential requirements towards achieving successful project execution. Project managers need reliable cost estimating systems to prepare their budgets and develop financial management plans. Cost estimation is simply a process of determining the cost and resources required to execute a project within a definite time frame. The use of BIM has proved to be more efficient and cost-effective than the traditional cost estimation methods. BIM is designed to input several project data where information such as cost, materials, and performance of the project can be accessed. The use of BIM makes it possible for the cost estimators to have access to directly generated takeoff, counts, and measurements consistent with the project design (Elbeltagi et al., 2014). This enables the project managers and team members to get quick and accurate cost estimates for quantifying materials and the cost effects of any adjustments made in project execution. These benefits have helped to minimize the shortcomings of the traditional cost estimation techniques, which take a long time and can potentially have mistakes due to human error. According to Abanda et al. (2017), some of the leading BIM soft wares packages include project-wise navigator, Autodesk QTO, Navisworks, CostX, Innovaya, ProjectWise Navigator, etc. Figure 3.11. Illustrates the process of cost estimation using these software packages.

3.2.6 Communication

Communication is a significant part of project management which entails the flow and dissemination of relevant information among project participants. Building Information Modeling with special emphasis on 'I' Information' has dramatically impacted the communication network throughout the project life cycle. BIM is designed mainly to communicate and share knowledge. This model provides an integrated communication and collaboration system between different stakeholders, including the architects, designers, engineers, clients, contractors, authorities, and even the end-users. BIM is a digital building model with descriptive information and object-oriented geometric, which coordinates data exchange between project participants by linking information through an information carrier to the various stakeholders (Melzner et al., 2015). The introduction of BIM for communication demand that projects managers and their team are equipped with digital tablets which enable them to get quick access to all the relevant information on the construction site. Through this digital process, the project teams can communicate and collaborate more accurately, thereby resulting in better quality and delivery on the site. Figures 3.12 and 3.13 depicts the communication and knowledge sharing among project participants with a BIM framework model.

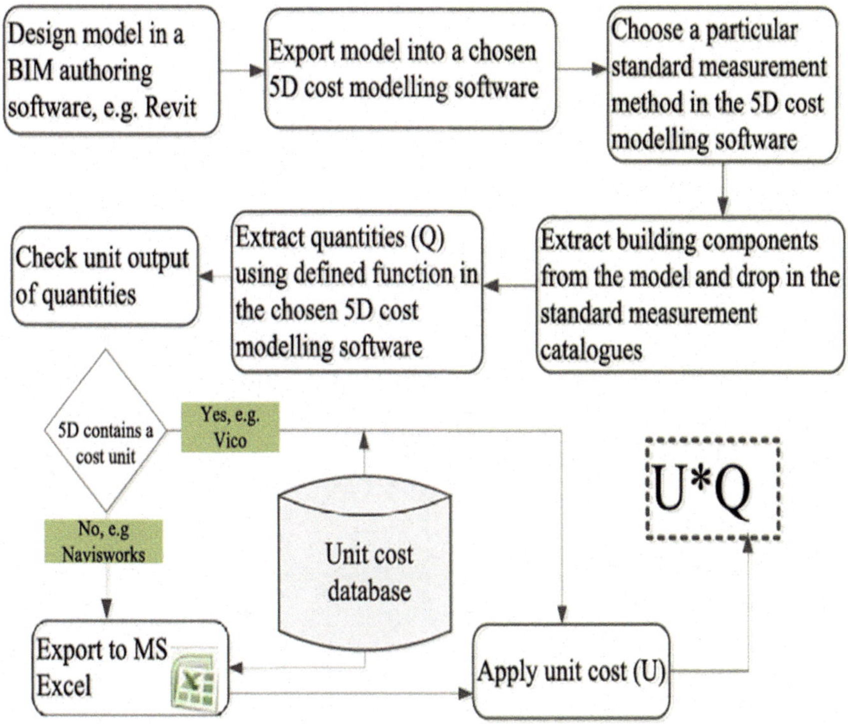

Fig. 3.11 Cost estimating process in a BIM-based cost estimating software (Abanda et al., 2017)

3.3 Case Studies of Building Information Modeling Application in Construction and Resultant Benefits

3.3.1 Case Study One: Venetian Casino Construction in Macau, Hong Kong

A Venetian Casino Construction case in Macau involves a podium façade, with a design reflecting Balinese styles managed by Hsin Chong Mace Management. The design phase reveals some complex coordination issues between the architect and structural engineers. This led to the adoption of the BIM 3D model, which detects queries such as conflicting information between architectural plans, sections, and elevations, misalignment, and setting out issues between the structural framing and GRC panels. The use of the models throughout the design coordination process facilitated better understanding between the client, architect, and engineers. At the end of the scheduled 12-weeks tender documentation, a detailed interactive model of the podium façade was presented. As the client noted, the BIM coordinated tender

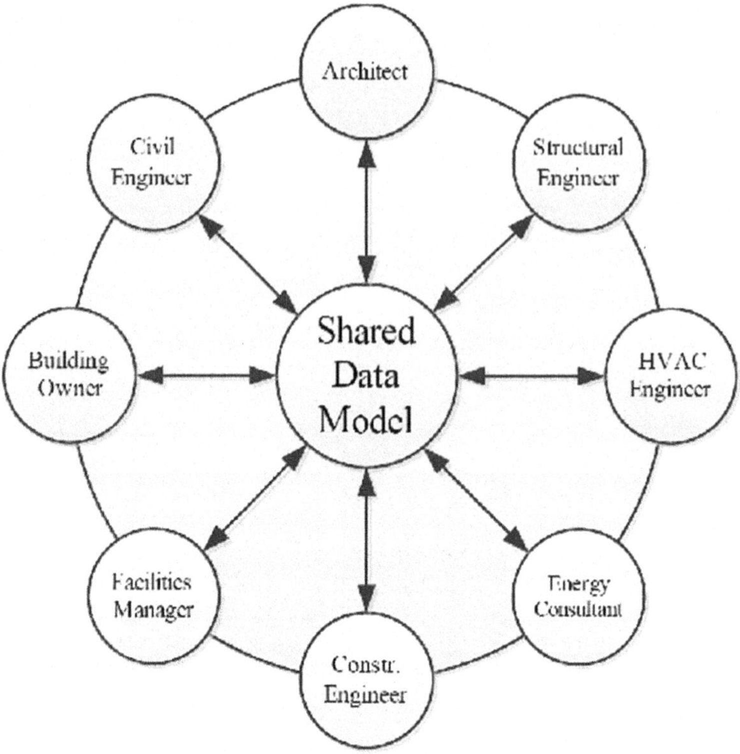

Fig. 3.12 Building information modeling for sharing and exchange of information (Liu et al., 2015)

was completed within a shorter period than the previous phase (Rowlinson et al., 2010).

3.3.2 Case Study 2: Cathay Pacific Cargo Terminal, Hong Kong

A Cathay Pacific Cargo Terminal case situated at the Southside of the Hong Kong International Airport platform costing roughly USD 500 M is presumed to be the largest air cargo terminal in the world. At peak times, the cargo terminal is expected to process more than 75 flights per day. The client specified the design and coordination of the project using BIM. BIM was further adopted to incorporate the project design, drawings, specifications, and a detailed clash detection analysis matrix in the model. In the BIM process, clashes were identified using the specialized clash matrix, which helped to integrate the cargo's design specifications and operational requirements.

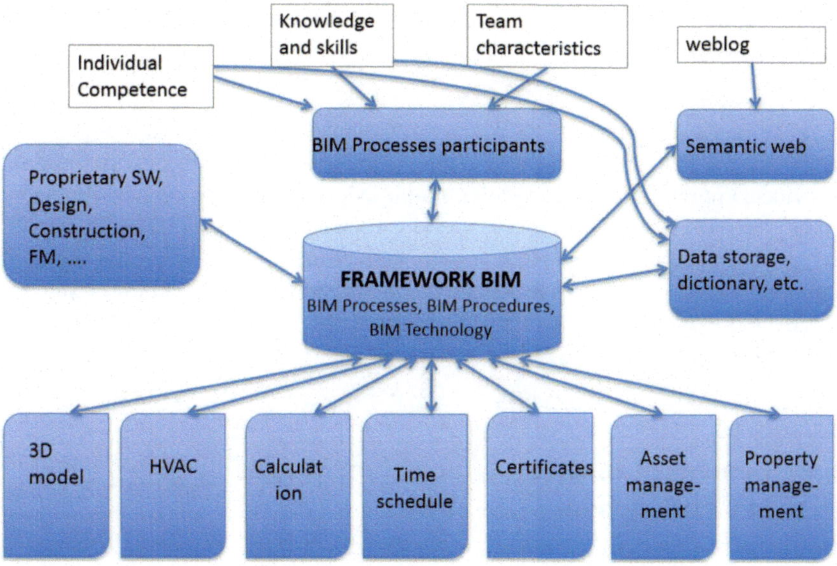

Fig. 3.13 BIM framework and communication flows (Nyvlt & Novotny, 2019)

This resulted in lesser cost over-runs, delays, on-site RFIs, smoother operation, and better collaboration among team members (Rowlinson et al., 2010).

3.3.3 Case Study 3: Aquarium Hilton Garden Inn, Atlanta, Georgia

The Aquarium Hilton Garden Inn project comprised a mixed-use hotel, retail shops, a proposed budget of Project $46 million, and a 484,000 square-foot hotel and parking structure. The project was designed using BIM to develop the architectural, structural and mechanical, electrical, and plumbing models of the proposed facility. These models were used for clash detection and potential collision between various structural and mechanical systems, visualization, work sequencing, collaboration among the team installation by subcontractors. As a result, 55 clashes were identified, resulting in cost avoidance of $124,500, cost–benefit of $200,000 and 1, 143 h saved (Azhar, 2015).

3.3.4 *Case Study 4: National Cancer Institute (NCI) in Malaysia*

The National Cancer Institute project is a government project with 10.65 acres of land, a 690 million RM budget, and a schedule of 36 months. The project was proposed to provide comprehensive care to patients with special emphasis on cancer treatment in Malaysia. BIM was adopted for 3D visualization of project design, resolve fabricator issues, and identify clashes at an earlier stage of the project, which helps close the communication gap among team members. As a result of this, 1800 clashes were discovered, which help minimize construction costs, time and improve the quality of the project (Azhar, 2015).

3.4 Benefits of Building Information Models in Construction

The introduction of Building Information Modelling offers several benefits to the stakeholders and the construction industry. Below are some of the benefits:

(a) **Capture Reality**: Reality is a representation of a subject from the real world. BIM provides/shows a model representation of a building, site, or project that allows the participants involved in a project to capture accurately and quickly all data related to the project. This has brought about innovation from the traditional methods using manual blueprint paper by the architect and engineers, which is often difficult, time-consuming, and inaccurate.

(b) **Enhance Information Flow**: The BIM provides an integrated construction system that permits the professionals such as architects, engineers, and project managers involved in the project execution to access all information relating to the project from anywhere. This allows each participant to subtract, add, and change information for the successful performance of their task (Boukara & Naamane, 2015). Consequently, the flow of information among professionals improves productivity and reduces error.

(c) **Time Effective**: The traditional method of capturing the sites takes a long time to collect information required by project participants to execute the project. BIM provides a faster and effective process of collecting, sharing, and using information which speeds up the design and the construction process. It saves time in the entire designing process where the workflow and drawing are produced directly from a single unified model. For instance, the BIM drawing device performs faster than the 2D drawing tools due to its connectivity to a quick and automatically updated database.

(d) **Improve Collaboration**: Collaboration is vital to the success of construction projects. The project participants have recognized the sharing of information and knowledge as a vital component of a successful project. Building information models is a collaborative system between construction participants, making information sharing easier than traditional drawing tools. Communication becomes better through the opportunity given to all team members to have input throughout the whole construction process.

(e) **Cost Reduction**: BIM provides detailed and better information of the building than a drawing set which helps to minimize the incidence of errors, thereby preventing rework and reducing cost. Also, the model makes it easy to adjust materials without incurring additional costs. As a result, better cost estimation can be achieved with the benefit of high precision of information through BIM use.

(f) **Conflict Resolution**: The adoption of BIM provides solutions for resolving conflicts that arise during project execution. This is achieved by early detection of potential errors that might occur on the project and necessary adjustments before the commencement of the project.

(g) **Better Visualization**: Effective visualization is vital to several activities conducted in the Architecture, Engineering, and Construction (AEC) industry (Mikael et al., 2015). One of the benefits of BIM is the ability to visualize a building from a different environment. It helps the participants to visualize what is intended to be built at any time of the project and in a virtual setting, thereby detecting any possible problem associated with the design and the construction process. As a result of this function, there is a reduction of cost and better management of the project.

(h) **Improve Safety**: The adoption of BIM improves safety management in construction. Through this technology, workers are well informed of the working conditions, potential risks associated with their task before the commencement of the project. This helps the workers to take the necessary precaution and work efficiently. In line with safety benefits, BIM provides the clients and operators with a better understanding of the design, which allows adjusting the design at will to achieve the desired result, including crowd safety.

(i) **Improved relationship between clients and contractors**: A good relationship between contractors and clients is essential in achieving success in any project. BIM is a digital model that creates transparency where the contractors and clients can access all the model information. Through this model, the clients become a part of the project team and update and accurately visualize the project status from time to time. Hence, a good relationship is enhanced based on the assurance that the project aligns with the objectives and interests of the client.

(j) **Improve Performance**: Through BIM adoption, the ultimate benefit that accrues to a construction company, project team, and the industry is improved project performance. This is in terms of cost, time, safety, and productivity. BIM models are designed with several facilities and functions that help reduce cost, time, detect risk early, improve safety, promote communication and collaboration among project stakeholders. The sum of all these benefits enhances performance and productivity.

3.5 Risk of Adopting Building Information Modelling

There are several risks associated with the adoption of BIM. Since BIM use is expected to increase in the future, addressing the risk becomes essential for effective utilization of the technology. According to Ya'cob, et al. (2018), the risks can be categorized into four categories:

(i) **Technological Risk**: This refers to technical problems in applying BIM to a construction project. Majorly this risk emanates from the structure of BIM, which depends on a single information supply that meets the requirements of all the professionals such as architect mechanical, electrical, and structural engineers working on a project. Members of these project teams are expected to utilize the same building information model to carry out their tasks resulting in interoperability, complication, and compatibility., As a result, several other technical risks such as data sharing, data loss during file exchange, upgrading of technology, version control, and software compatibility are also presented under technical risk.

(ii) **Financial Risk**: Building Information Modelling requires a high cost of implementation. These include investment in software, hardware, acquiring BIM expertise, licenses and certifications, and training personnel on the use of technology. Also, the high cost of BIM implementation includes updating the software and modifying the work process and workflow.

(iii) **Management Risk**: The newness of the building information model presents some managerial threats. These include resistance to change from team members, traditional leadership style, the inadequate commitment of top management, shortage of experts of BIM, lack of knowledge of BIM application, lack of training, and low demand of BIM.

(iv) **Legal Risk**: The legal risk associated with BIM implementation emanate from the collaborative design of BIM, which integrates all stakeholders' roles and responsibilities on a project. s BIM procedures lack a standard policy, thereby creating a high risk of intellectual property. This makes it difficult to determine the data's ownership and assign responsibility to any project participant for any error detected. Invariably this can lead to disagreement among the project team.

3.6 Relevant Skills Required for Building Information Modelling

Building information modelling is premised on integrating skills from different backgrounds of the built environment for efficiency and project success. These sets of skills are essential for building, utilizing, and implementing BIM irrespective of the dimensions. These inevitable skills for BIM specialists can be grouped into two categories (Davies et al., 2015; Rahman et al., 2017). These include the following.

3.6.1 Soft Skills

These are skills that must be inherent in BIM teams, reflecting their personal traits, interpersonal management abilities, and individual differences. These skills must be possessed in moderate proportions and a balancing mechanism for achieving BIM objectives considering the disciplines of the actors. These soft skills include the following:

(i) **Communication Skill**: BIM is an information-based model that demands a high level of communication among the stakeholders. As a result, communication skill involves communicating unambiguously, and accurately with the stakeholders of the project.

(ii) **Interpersonal Skill**: BIM is a collaborative process that requires the support of various units towards the successful accomplishment of the project. Interpersonal skills involve the ability to tolerate and respect personal views, opinions, and professional resolutions regarding the BIM operations. This skill is necessary for BIM operators in enhancing the quick resolution of clashes between models.

(iii) **Negotiation and Conflict Management Skill**: The collaborative nature of projects and clash detection give rise to conflicts. BIM helps to detect clashes early before the commencement of the project. Accordingly, negotiation and conflict management skills are essential for BIM operators to quickly resolve disagreements and prevent clashes from overriding and diminishing BIM project objectives.

(iv) **Team Playing Skill**: Team playing is the ability to contribute effectively to individual roles, especially cautiously looking out to control others' poor performance. BIM demands input from diverse expertise on the project team to achieve its objective. This is important as the knowledge of each member is highly valuable for the process of BIM application. Hence, it becomes necessary for members to possess a team playing skill to develop the best strategy for successfully implementing a BIM project.

(v) **Leadership Skill**: This is the ability to queue behind the team leaders' directives, instructions, and visions. Leadership skill is a necessary soft skill required in the BIM environment. Each team member must have control over their work and contributions and protect their information and process. This is important to ensure successful implementation and compliance with the BIM agreement.

(vi) **Planning and organizations**: The application of BIM entails developing activities and processes required to achieve a quality project that is cost-effective, within the scheduled time, and meets the client specification. As a result, BIM entails the ability of the operators to plan and develop procedures, organize activities and determine the workflow to achieve the objectivities.

(vii) **Analytical Skill**: BIM is a digital model that draws a vast amount of data and information across a project's life cycle. This takes place from the designing through the construction process to the maintenance of the project. Analytical skills entail the ability to collect and analyze information/data to make the most effective decision. This skill is vital in the BIM environment as it helps in decision-making. Hence, BIM operators require analytical skills to unleash the full potential from their building information models.

(viii) **Problem Solving Skill**: BIM demands the ability to use independent judgment, problem-solving, and priorities-based factors. A problem-solving skill involves defining a problem, determine the cause of the problem, identify alternatives course of action and implementing the best solution

3.6.2 Technical skills

These skills are based on knowledge, understanding, and utilizing BIM software, tools, and technologies.

(i) **Information Technology /Computer Literacy**: The primary technical skills required of BIM operators are computer literacy and information technology (IT) proficiency.

(ii) **Proficiency in BIM Software**: The software skills are fundamental competencies required for BIM operators and professionals. Examples of this software include the AutoCAD, Revit, SketchUp, AutoCAD MEP, ARCHICAD, Navisworks, Vector works Architects, SolidWorks Electrical 3D, Civil 3D, Revit Live, Data CAD, Vector works Designers, MicroStation, Revit LY, BIM Track, Vectorworks landmarks, Leica Geosystems, Trimble SysQue, Aurora, IrisVR, Matterport, The Wild, Autodesk Ecotest Analysis, Iconstruct; ALLPLAN, AECOsim Building designer, cadmatic 3D Plant design, GT Digital project, Intergraph Smart Fabrication, planning & execution, Tekla structural designer, ARCADia BIM 11, Edificius, Green building studio, GT STRUDL, and Primavera.

3.7 Summary

Building Information Modeling has rapidly become an indispensable technology in the construction industry. Globally, this technology has proved to be data-rich intelligent models through which a virtual prototype of a building design and the construction process can be developed before commencing a project. Therefore, making it easier, cheaper, and effective. Over the years, BIM has evolved into various dimensions ranging from the 3D geometry, 4D schedule, 5D Model cost estimation analysis, 6D sustainability, 7D facility management, and 8D occupational safety and hazards model. These models have improved the data processing by incorporating information required for better comprehension of project activities. As a result of BIM application, the industry has experienced a commendable improvement in design visualization, constructability analysis, site planning and utilization, scheduling, sequencing, cost estimation, communication. Case studies of BIM application in the construction process have shown several benefits: clearer visualization, early detection of conflicts, better communication, and collaboration among stakeholders, safety, fewer cost overruns, and timely delivery. These benefits have enhanced the overall quality of projects and the performance of the construction industry. However, BIM adoption also presents some risks, which can be categorized into technical, financial, legal, and managerial risks. Examples of such risk include data sharing, high cost of implementation, shortage of BIM experts, and intellectual property. Finally, the performance of BIM demands specific skills requirements. Both soft and technical skills are necessary for successful utilization of the technology to deliver quality projects. This presents an excellent challenge for the industry to train and equip the construction employees with the required skills.

References

Abanda, F., Foguem, K., & Tah, J. (2017). BIM – New rules of measurement ontology for construction cost estimation. *Engineering Science and Technology, 20*(2), 443–459.

Associated General Contractors (AGC) (2005). Contractors' guide to BIM. Edition 1, Retrieved from: https://www.engr.psu.edu/ae/thesis/portfolios/2008/tjs288/Research/AGC_GuideToBIM.pdf

Azhar, S. (2015). Building information modelling (BIM): trends, benefits, risks, and challenges for the AEC industry. *Leadership and Management in Engineering, 11*(3), 241–252.

Bazjanac, V. (2006). Virtual building environments (VBE) – Applying information modelling to buildings. Retrieved from http://repositories.cdlib.org/lbn/LBNL-56072.

Barati, R., Charehzehi, A., & Preece, C. (2013). Enhancing planning and scheduling program by using benefits of BIM-based applications. *Civil and Environmental Research, 3*(5), 41–48.

BIM community (2019). 4D construction sequencing: An introduction. Retrieved from https://www.bimcommunity.com/events/load/608/BIM-2019.

Boukara, A., & Naamane, A. (2015). A brief introduction to building information modelling (BIM) and its interoperability with TRNSYS. *Journal of Renewable Energy and Sustainable Development, 30*, 126–130.

Davies, K., McMeel, D., & Wilkinson, S., (2015). Soft skills requirements in a BIM project team. In: Proceedings of the 32nd CIB W78 Conference 2015, 27 -29th October 2015, Eindhoven, Netherlands.

Elbeltagi, E., Hosny, O., Dawood, M., & Elhakeem, A. (2014). BIM-based cost estimation/monitoring for building construction. *International Journal of Engineering Research and Applications, 4*(7), 56–66.

Foster, L. L (2008). Legal issues and risks associated with Building Information Modeling Technology. ProQuest.

Forgues, D., Iordanova, I., Valdivesio, F., & Staub-french, S. (2012). Rethinking the cost estimating process through 5D BIM: A case study construction research congress. *American Society of Civil Engineer (ASCE), 2012*, 779–786.

Hijazi, W., Alkass, S., & Zayed, T. (2009). *Constructability assessment using BIM/4D CAD simulation model.* AACE International Transactions, Seattle, WA.

Hossan M., & Yeoh, J.K., (2018). BIM for existing buildings: Potential opportunities and Barriers. Conference paper: 3rd international conference on building materials and construction, at Nha Trang, Vietnam. In: *IOP Conferences Series: Material Science & Engineering*, vol. 371, 012051, pp. 1–9.

Kamardeen, I., (2010). 8D BIM Modeling for accident prevention through design. In: Egbu, C. (Ed) *Proceedings of 26thAnnual ARCOM Conference, 6–8 September 2010, Leeds, UK, Associations of Researches in construction Management*, pp. 281 –289.
Kumar, S.S & Cheng, J. (2015). A BIM-based automated site layout planning framework for congested construction sites. *Automation in Construction*, 59, 24–37.
Liu, X., Xie, & Jia, J (2015). WebVis BIM: Real-time Web3D visualization of big BIM data. In: *Proceedings of the 14thACM SIG Graph International Conference on Virtual Reality Continuum and Its Application in Industry*, pp. 43–50.
Melzner Feine, I., Hollermann, S., Rütz, J., & Bargstädt, H. (2015). The influence of building information modeling on the communication management of construction projects. In: *15th International Conference on Construction Applications of Virtual Reality, At Banff Canada*, 5-7 Oct 2015
Mesaros, P., Smetankova, J., & Mandicak (2019). The fifth dimension of BIM – Implementation survey. *Earth and Environmental Science*, *222*(01), 1–9
Mikael, J., Mattias, R., & Petra, B. (2015). Real-time visualization of building information models (BIM). *Automation in Construction, 54*(2015), 69–82.
Nyvlt, V., & Novotny, R. (2019). Sharing knowledge and information within BIM. *Life Cycle Processes, MATEC Web of Conferences on Building Defects, 279*(01001), 1–6.
Rahman, R., Tang, P. & Ayer, S. (2017). Building information modelling skills in construction-related disciplines: A social network and job advertisement-based comparative analysis 7(1)
Rightserve Solutions (2019).Advantages of 3D BIM Modelling. Retrieved from https://medium.com/@rightservekochi/advantages-of-3d-bim-modelling-F539067d5a81.
Rowlinson, S., Collins, R., Tuuli, M., & Jia, Y. (2010). Implementation of building information modeling (BIM) in construction: A comparative case study. In: *Proceedings of the 2nd International Symposium on Computational Mechanics and the 12th International Conference on the Enhancement and Promotion of Computational Methods in Engineering and Science*, Nov 30 - December 2, Hong Kong & Macao, 572–577.
Sai Evuri, G, & Amiri-Arshad, N. A. (2015). A study on risk and benefits of building information modelling (BIM) in a construction organization. Retrieved from http://hdl.handle.net/2077/39971.
Sulankivi, K., Makela, T., & Kivinjemi, M. (2009). BIM-Based Site Layout and Safety Planning, First International Conference on Improving Construction and Use Through Integrated Design Solutions: CIB IDS 2009 VTT Technical Research Centre of Finland, 10 June 2009–12 June 2009, 3(5) 125–140.
Weebly (2020). BIM facilities management. Retrieved from https://8dbim.weebly.com/6d.html.
Worden, K. (2016). *BIM and Communication: Implementation of Building Information Modelling into an Integrated Project Delivery Contract to Encourage Project Teams to Communicate, a Thesis of the Faculty of California Polytechnic State University, San Luis Obispo.*
Ya'cob, I. A., Rahim, A. Z., & Zaninon, N., (2018). Risk in implementing building information modelling (BIM) in Malaysia construction industry: A review. In: *E3S Web of Conference* 65; 03002(2018). ICCEE, 1–9.
Yang, H., Lee, M., Siao, F., & Lin., Y (2013). *Use of BIM for Constructability Analysis in Construction, Proceedings of the Thirteenth East Asia-Pacific Conference on Structural Engineering and Construction* (EASEC-13), September 11–13, 2013, Sapporo, Japan, A-3–1.
Yung, P., & Wang, X. (2014). A 6D CAD model for the automatic assessment of building. *International Journal of Advanced Robotic Systems Sustainability, 11*(01), 1–8.

Chapter 4
Drone/Unmanned Aerial Vehicles (UAVs) Technology

4.1 Introduction

Drone technology, also known as Unmanned Aerial Vehicles (UAVs), has digitally transformed the construction industry and gained much application in architecture, engineering, and construction operations. The digitalization of the construction industry, recently known as construction 4.0, has made drone technologies a piece of essential equipment for proper monitoring of construction activities from commencement to completion of the project.

Historically, drones have been around for some while now. However, before now, drones were mainly built and researched for military purposes. Over the years, and with tremendous improvements in drone technology, civilian applications of UAVs have gained wide popularity (Albeaino et al., 2019; Tatum & Liu, 2017; Zhou & Gheisari, 2018). It was noted by Sanson (2019) that in 1849, the Austrian empire utilized hot air balloons to drop bombs on Venice in the first Italian War of Independence. The first exemplar of modern UAV with the capability of an autonomous flight coupled with -range and wireless links to the ground station was developed in 1944 during the Second World War (Giordan et al., 2020). About the year 2000, the recreational and commercial drones began to show up and by 2006 issuing commercial drone permits in the United States started, with about two permits issued per year (Devers, 2019; Ford, 2018). Thereafter, drones such as Parrot AR Drone and DJI Platinum were available for civil and commercial purposes (Devers, 2019; Lin, 2018). In 2014, the development of drone technology for commercial use caught the construction industry's attention (Lin, 2018). Since 2014, the construction industry has experienced several applications of drone technologies in areas such as: site security, safety inspection, construction monitoring, etc.

Drones technology can be described as a flying robot that can be remotely regulated by flight software and Global Positioning systems. A drone is a recognizable airplane remotely controlled for data collection. Drones vary in size, and it contains topographic measurement used to survey buildings, roads, highways, bridges and any surface region. These technologies are built with intelligent stabilization systems,

O. Adepoju et al., *Re-skilling Human Resources for Construction 4.0*, Springer Tracts in Civil Engineering, https://doi.org/10.1007/978-3-030-85973-2_4

sensors, and cameras that can fly, perform dedicated functions, and obtain high-quality video photography (Accenture, 2016). They are structured to provide information regarding work progress, locations of certain objects on larger sites, and even survey elevations and topography of the prebuilt land (Higgins, 2017). Generally, the term drone refers to autonomously guided aircraft, including submarines or land-based autonomous vehicles (Yaacoub et al., 2020).

Drones are likened to an automated plane and are often called Unmanned Aerial Vehicles (UAVs); Unpiloted Aircraft System (UAS); Remote Piloted Aircraft System (RPAS); Quadcopter etc. (Albeaino et al., 2019; Fox, 2017; Giordan et al., 2020; Tatum & Liu, 2017). Generally, drones can be described as a device capable of flying and can either be controlled remotely or controlled autonomously by software. Drones mainly focuse on "flying" and are controlled remotely or autonomously. Unmanned Aerial Vehicles (UAVs) do not require a pilot, but are airborne systems controlled through ground control stations. Tatum and Liu (2017) defined an Unmanned Aircraft System (UAS) as an aircraft system that a pilot on ground flies. Grayson (2016), as cited in Tatum and Liu (2017), noted that the term drone might also broadly refer to any equipment operated independently of human control. Giordan et al, (2020) also stated in their study that a UAV is referred to as an unmanned aerial system, which can have an autonomous flight with or without an engine, can be remotely controlled, and collect data. Defining a construction drone would be tasking, but it can be defined as a device that is airborne, controlled remotely, or through software deployed for construction activities: such as traffic monitoring, surveillance or survey mapping, site navigations, material tracking, and inspection, etc.

In recent times, the use of drones in the engineering and construction industry is increasing. It has enhanced the key stages of delivering a quality project, including pre-planning, planning, inspection, construction and post-construction, and monitoring. Drones have brought about developments in the construction industry by improving safety, reducing cost, reporting precision, and increasing efficiency. Construction projects are rapidly evolving beyond past standards, and commercial drones are driving technical evolution, thereby spurring industrial change across construction sites (DeYoung, 2018). In the construction industry, drones have proven to be a valuable technological development due to their success in data sharing for project execution. Also, drones present aviation opportunities and abilities that have helped resolve many challenges faced on the construction site. In the construction industry, drones are used explicitly on construction sites for several purposes. These include project design pre-construction site assessment, mapping, aerial surveying, material stockpiles, locating utilities, workers and equipment measurement of excavation pits, monitoring site progress, inspection, security, and safety. Drones are equipped with high-resolution cameras, Global Positioning System (GPS), ultrasound devices, radio-frequency identification readers, laser scanners, and remote sensors for: imagery, data collection to detect objects, surveying, measure distance, inventory, and equipment tracking. Today, the industry has experienced tremendous improvement due to drones and UAVs used for several stages of the construction projects. In light of the growing advancement in drones, drones are expected to create more jobs, thereby increasing drone application usage in the construction industry. Drones are manufactured for commercial and combat purposes. Military drones are the first innovation of drones specifically manufactured for defense and used by military force for intelligence surveillance and weapon of delivery during combat.

On the other hand, commercial drones are drones manufactured for the use and control of civilians. For many years now, commercial drones are primarily used in the construction industry (Zitzman, 2018). According to the Federal Aviation Administration (2018), approximately 79% of commercial drone usage is used for public and private construction projects, aiding project site inspections and civil engineering endeavors.

4.1.1 Components of a Drone

A drone can be sophisticated based on components, uses, and area of application. In its simplest form, the main elements of a drone are classified into three which are;

i. the aerial platform (for flying);
iii. the ground control station (for remote or autonomous control); and
iii. the communication system (for synergy between components) (Giordan, 2020).

Components of a drone are based on authors and expert lexical register; There are several classifications of the elements of drones provided by various authors. Tatum and Liu (2017) stated that the elements of drones include the drone itself and the control system, ground and satellite-based equipment, communication links, and operators which are required to operate the aircraft effectively and safely. Sanson (2019) itemized components of a typical drone as consisting of a: quadcopter, four propellers, motor, landing gear, battery, camera, and remote control.

Tatale, Anekar, Phatak, and Sarkale, (2018) also stated that the main part of the quadcopter is a frame that has four arms and should be light and rigid to host: a LIPO battery, four brushless DC motors (BLDC), controller board, four propellers, a video camera and different types of sensors along with a light frame as shown in Fig. 4.1.

According to Ostojić, Stankovski, Tejić, Đukić, and Tegeltija, (2015), quadcopters can have a very different hardware component which is dependent on the application in which they will be employed. They furthered that the standard features of a quadcopter are microcontroller, sensors, motors, Global Positioning System (GPS) power supply, and telemetry devices.

In summary, the component of a drone may vary based on the control system, either autonomous using programmed navigation software or remotely controlled by a person). Regardless of the variations, components that can make a drone fly, communicate with the base station (that is, operate autonomously or gather data into a storage device), and onboard communication between two or more components must be available. Considering the various components enumerated by multiple authors, a summary of the various components of a drone and their functions is detailed below.

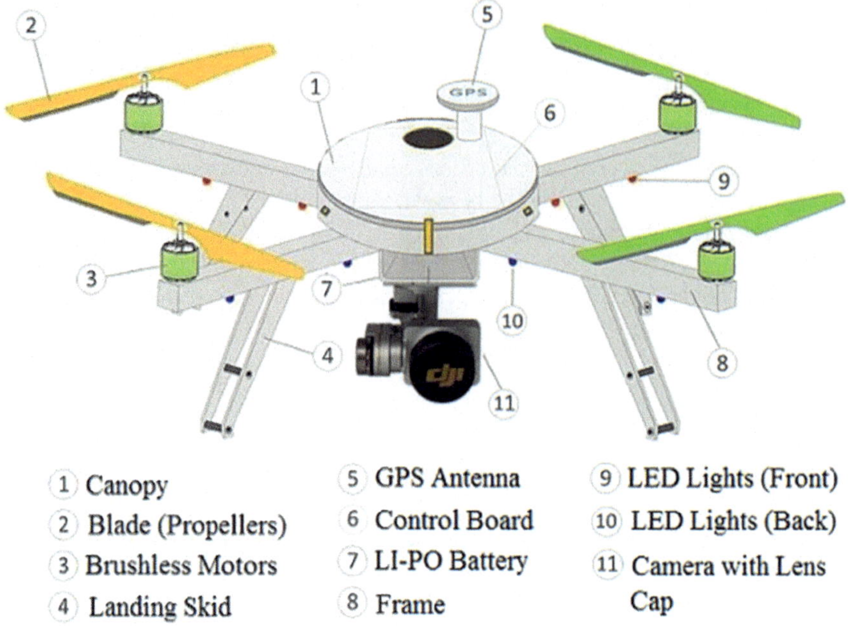

Fig. 4.1 Drone/quadcopter structure (Tatale et al., 2018)

Components that enable a drone to fly include the following:

a. **Battery:** all electronics need electricity to function. Therefore, the battery supplies the necessary electricity required for the drone and its components to work. In light of the preceding, the battery can be referred to as the life of the drone—once it goes off, the drone comes down. The higher the battery capacity, the more time a drone can stay fully functional in the air.
b. **Propeller and Motor**: These are used for navigation and take-offs. The propeller and motor provide thrust, which are responsible for the drone's movement from the ground up and back on the ground. They are the wheels of the drone.
c. **Frame**: This allows for the attachment of other devices. The frame is the body of the drone and also houses the weight of the drone.
d. **Landing Gear**: This is the stand or leg of the drone, and it is important for balance while on the ground or in the air. The landing gear bears the weight of the frame and other components attached to the frame.

Components that establish communication with a base station and visuals include the following:

a. **GPS Antenna**: This component is used to establish a proper connection between the drone, location, and base station for adequate interaction. The GPS is responsible for sending and receiving tracking, location, mapping, location signals, etc. The GPS and other sensors like infra-red and ultrasonic sensors are used for navigation purposes, to avoid collision, and altitude control.

b. **Camera**: This is used for visuals when flying. The camera is the essence of the drone because the work of a drone is to get you where you cannot access. The camera helps in gathering programmed data by taking aerial snapshots, providing imagery and video recording to the mapped area.

Components that establish communication between two or more components include the following:

a. **Control Board**: This is the drone's brain, and it helps coordinate, give and interpret instructions and commands given to a drone. The control board is responsible for telling other components what to do, when to do it and how to do it when a remote user or programmed software initiates a command.

b. **Microcontrollers**: these provide a set of digital and analog I/O pins that can be interfaced to various expansion boards (shields) and other circuits. The microcontrollers control the overall performance of a quadcopter, such as flying mechanisms and live streaming of videos. The boards feature serial communications interfaces, including USB on some models for loading programs from personal computers (Tatale et al., 2018).

4.1.2 Classification of Drones

Classification of drones varies, but basically, drones are classified as autonomous and non-autonomous depending on the control system (Kazi et al., 2018). Autonomous drones are controlled from remotely spaced programmable computers using an onboard autopilot and global positioning system (GPS); in other words, an autonomous drone flies and carry out programmed functions without human interference. Non- Autonomous drones are controlled or piloted manually using a remote controller requiring a skilled operator. Structurally, a drone can also be classified as an Aircraft drone or a Copter drone (Kazi etal., 2018). Aircraft drones are fashioned to look like planes with wings attached to it for flight; these types of aircraft are popularly used by military and security operatives and are famous for their high-speed flying ability. An aircraft drone needs a runway to take off and land. Copter drones are fashioned after a rotorcraft, or like a helicopter whose flight and thrust are supplied by spinning rotors. Drones within this classification are often referred to as quadcopters and are suitable for recreational and commercial purposes within a highly dense urban area (Kazi et al., 2018).

Unmanned Aerial Vehicles are multifaceted, with over 28 variations based on the single qualifier of flying principles alone. The most inclusive classification of Drones is based on operating mass/weight and UAV's purpose of flight/usage. Classification based on operating mass can either be heavier-than-air or lighter-than-air. On the other hand, the classification of UAV based on flight usage is categorized

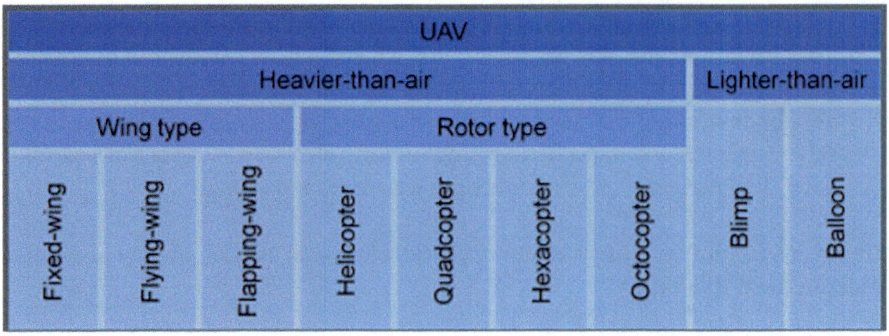

Fig. 4.2 Common unmanned aerial vehicles classification (Liew et al., 2017)

as flight time versus UAV mass and degree of autonomy versus degree of sociability. Figure 4.2 illustrates common UAV types given a respective vehicular mass differentiation between heavier-than-air and lighter-than-air.

4.1.2.1 Classification of Drones/Unmanned Aerial Vehicles Based on Operating Mass/Weight

(i) **Heavier-than-air:** The heavier-than-air UAVs have substantial vehicle mass relying on aerodynamics and propulsive thrust to fly. This classification of drones is sub-classified into *wing* or *rotor* types for drone type determination. Wing-type UAVs comprise fixed-wing, flying-wing, and flapping-wing methodologies to generate aerodynamic lift. On the other hand, Rotor-type UAVs include a plethora of multi-rotors by combining multiple rotors and propellers, which point upwards to create propulsive thrust for its lift (Fig. 4.2).

(ii) **Lighter-than-air:** The lighter-than-air UAVs rely on buoyancy forces such as helium gas or heated air to fly. They have an extended stamina flight capacity which can be used for photography and surveillance. One major **setback/constraint** of the lighter-than-air UAVs is its inability to maneuver, **requiring tethered/rope** to keep it from drifting due to the wind. Examples include blimps and balloons (Fig. 4.3).

4.1.2.2 Classification of Drones/Unmanned Aerial Vehicles Based on Purpose of Flight/Usage.

(i) **Flight time versus UAV Mass**: One factor to consider when purchasing a drone is flight time. On average, the flight time of drones ranges from 15 to 30 min, dependent on the battery energy, design range of flight, and weather condition. However, a non-racing drone can fly between 30 to 60 min, while a racing drone can fly for 125 min per hour. More often, the battery capacity

Fig. 4.3 Categories of Heavier-than-air drones (Hassanalian & Abdelkefi, 2017)

determines the flight time of a drone. Also, UAV mass implies the weight of drones which also varies. It can be bigger/heavier or lighter and smaller.

(ii) **Degree of Autonomy Versus Degree of Sociability**: The degree of autonomy is the capability of drones or UAVs to take off, carry out the specific task and return to their base without significant human involvement. In essence, the technology is programmed with the tasks required to be executed without the need of an operator. The attainment of UAV autonomy shows a significant level of achievement of any technology inventor. The degree of sociability describes the extent to which a drone can interact or collaborate with humans. Such categories of drones are developed with software that enables them to communicate with a human. As a result, these drones have a lower degree of autonomy due to manual control. (Liew et al., 2017).

4.1.3 Types of Drones Used in the Construction Industry

In recent years, drones in the construction industry are one of the most striking trends. This accounts for almost a 240% increase in drone use compared to other sectors of the economy. Even though there are various types of drones, the construction industry mainly uses commercial drones. Examples of these drones are listed below (Tkac and Mesaros (2019)

(i) **Single Rotor**: A single rotor is a type of drone one rotor with a propeller to control it. The single rotor drone comes with a simple design, is low cost, is easy to start and move, and has shown greater efficiency than the drones with multiple rotors (multi-rotor). However, they are nor no strong, which requires caution when landing and a lot of maintenance

(ii) **Multi-Rotor**: Multi-rotor is a common type of UAV that makes use of more than two rotors. They are categorized based on the number of rotors. For example, the UAV's that make use of three rotors are called the tricopters, while four rotors are called quadcopters, six rotors hexacopters, and eight rotors UAV's are called (octocopters). Multi-rotors comprise fixed-pinch spinning blades that produce lift. Multi-rotor drones are used for professional aerial mapping and pleasure. These types of drones are primarily used for video recording, photography, and aerial surveying. Two significant advantages of motor rotors are the drone's ease of operation and affordability option in getting "an eye in the sky." However, multi-rotor drones are limited in speed and endurance. As a result, it cannot be used for long-distance inspection, large scale aerial mapping, and long endurance monitoring. An example of a multi rotor drones is the DJI Matrice 210 used at the Allianz Fieldin Minnesota (Fig. 4.4).

Fig. 4.4 An example of a multi rotor drone (Gibson & McManmon, 2018)

(iii) **Fixed Wing Drones**: A fixed-wing drone which can also be referred to as an-unmanned aerial vehicle or system, is a firm structure of aircraft that used a fixed or static wing to generate lift. It is custom-built with an internal engine to ensure proper functioning. This Fixed-wing drones can fly longer distances within a short time. As a result, it is considered more efficient than other types of drones regarding topographic mapping, energy efficiency, and its ability to reach a high altitude during flight. Also, this type of drone has a higher carriage capacity (payloads) and survives in harsh weather conditions. The fixed-wing drones also present some shortcomings regarding landing as it is restricted to moving forward only. In time past, this type of drone was used by the military for surveillance during the war. Nevertheless, in recent times it has become a standard tool used by several industries, including construction. Common fixed-wing drones include straight wing, delta wing, and a swept wing.

(iv) **Hybrid Systems**: Hybrid systems of drones are drones developed to solve some of the constraints of the single, multi-rotor, and fixed wings drones. This system of drones possesses the features of multi-rotors and fixed wings. This type of drone has wings and multiple rotors. These features enable it to take off quickly, climb, hover, fly a longer distance, and land with little energy consumption. It can hover like a helicopter and fly like a plane. It has the advantages of stability, flexibility, and fuel efficiency. Hybrid drones operate by flying a pre-schedule flight route at a specified height of a user, gather data through its color and multi-specified sensors, and land vertically to the starting point (Tkac & Mesaros, 2019). An example of a hybrid system of drones is the quadrotor (Fig. 4.5).

Fig. 4.5 A mini Quadrotor (Papa et al., 2017)

4.2 Application Drone Technology in the Construction Industry

Implementing drones' technology in the construction industry has made the construction work more efficient and sustainable. The use of drones has gained more attention in the construction and engineering field due to the swift, safe, and low-cost access to visual data, thereby providing spatial information about hazardous geographical locations. These technologies permit construction companies to utilize the data collected from drones to perform processes and improve almost all aspects of construction management. Consequently, drones/aerial vehicles are adopted in the construction industry for several purposes, which include the following:

I. Construction Site Inspections.

Inspection is a key activity in the construction process, and drone technology has enhanced this procedure. Drones are developed to collect visual data required for inspection, detect violations and track site progress. The use of drones for construction and inspection helps identify any form of technical problems on the constructions sites, which is important for delivering a quality project. Over time, using a drone has aided construction firms to work more effectively in managing resources, time and minimizing delays. Also, using drones for construction site inspections protects the workers from dangerous locations, thereby minimizing health and safety risks. Recent advancement in technology has made it possible to use a drone to fly around construction sites, to view things on the top of the roof of a skyscraper under construction and helped in the creation of a detailed 3D models of new construction projects (Young, 2019). To achieve an excellent result on construction sites, drones can be undertaken regularly and can cover larger areas (Ayemba, 2019) (Fig. 4.6).

II. Topography Mapping and Land Surveying.

The introduction and advancements in drones' capabilities have benefitted the construction industry with high-quality aerial mapping. Topography mapping using the drone is a process by which land is surveyed using Unmanned Aerial Vehicles. Maps, which are used to update clients on their project's progress, are easily accessible with the data collected by drone. These data collected via drone give an accurate image of the site required to properly plan a project. This information guides the contractors on the area to dig, fill and construct the building. The drone mapping process with the generation of high-quality aerial images often results in a 3D representation of the surveyed area. Consequently, the work of the surveyors is made easier and effective. Figure 4.7 shows how a drone can be used to switch maps into 3D for analysis and project planning and drone survey of acres of land in few minutes.

III. Laser Scanning and Aerial Photogrammetry.

Advancement in technology such as drones is developed with software facilities that can convert data collected from laser scanners and photogrammetry into: geographic,

Fig. 4.6 Construction Site Inspection by Drone Technology (Zitzman, 2018)

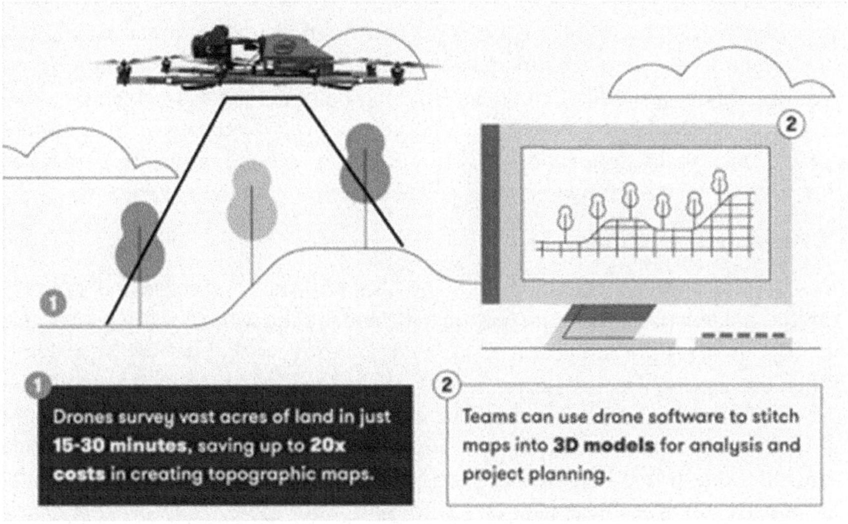

Fig. 4.7 Drone for Topographic Mapping and Land Surveys (Zitzman, 2018)

photographic, thermographic, and metric information required by construction professionals to design. Also, drones are developed with software facilities that can survey, and deliver quality projects. With the laser scanner facility of a drone, it becomes possible to measure a region of space following the predefined density. Also, drones are used for topographic, geological, and orthophotos surveys on the construction sites can achieve photogrammetry. Some of the photogrammetry

Fig. 4.8 Drone Application for Connecting Terrestrial Laser Scanning and Aerial Photogrammetry (Tkáč & Mésároš, 2019)

services provided by drones required for construction include processing contour lines, creating 3D maps and models, mapping construction sites monitoring with hydrogeological instability, and reconstructing point clouds. Consequently, drones have proffer solutions to some of the constraints faced by surveyors in accessing a suitable laser scanning location. Figure 4.8 depicts a building where a drone was used to achieve terrestrial laser scans and aerial photogrammetry (Fig. 4.8).

IV. Building Inspection.

In recent times, drone technology is being used to inspect buildings and structures. This has made it possible for inspection to be undertaken while construction is going on. This involves the inspection of the roof and areas within the structure that are difficult to access. The roof inspection is very vital to evaluate the current conditions and any defect that needs attention. The use of drones serves as a better option than the traditional method that requires the service of inspection providers/professionals. Generally, due to the height in reaching the roof area, the conventional methods that involve the use of ladders, or scaffolding, often pose some dangers and are time-consuming.

V. Automation and Data Collection.

Construction 4.0 has brought about the need and importance of automation and data collection for the prompt delivery of quality projects in the construction industry. The construction industry represents one of the sectors in the economy with significant adopters of aerial data. The use of drones makes it possible for construction tasks to be automated and conducted in correspondence with procedures. In construction, drones are applied to collect data required by the project team to execute the project successfully. This avails the team the opportunity to review and compare information for a better decision. In recent times, drones are structured for automated operations, which provides essential information required for construction. These include topography studies, design data, analysis of stockpiles, and cutting and filling parameters.

VI. Remote Monitoring and Progress Reports.

Drones are also referred to as Remote Piloted Aircraft System (RPAS), which encompass a flying platform remotely directed by a pilot to view and capture the entire construction site. One of the utmost benefits of drones in the construction industry is the opportunity for the project contractors and clients to view the construction site or ongoing project from any location. The capability of the remote monitoring of drones can be used to evaluate the progress of the construction work at different stages of implementation. This is helpful to the project manager, contractor, and clients as precise information on the project's day-to-day progress is provided. The overall components of drones for smart construction monitoring are described in Fig. 4.9.

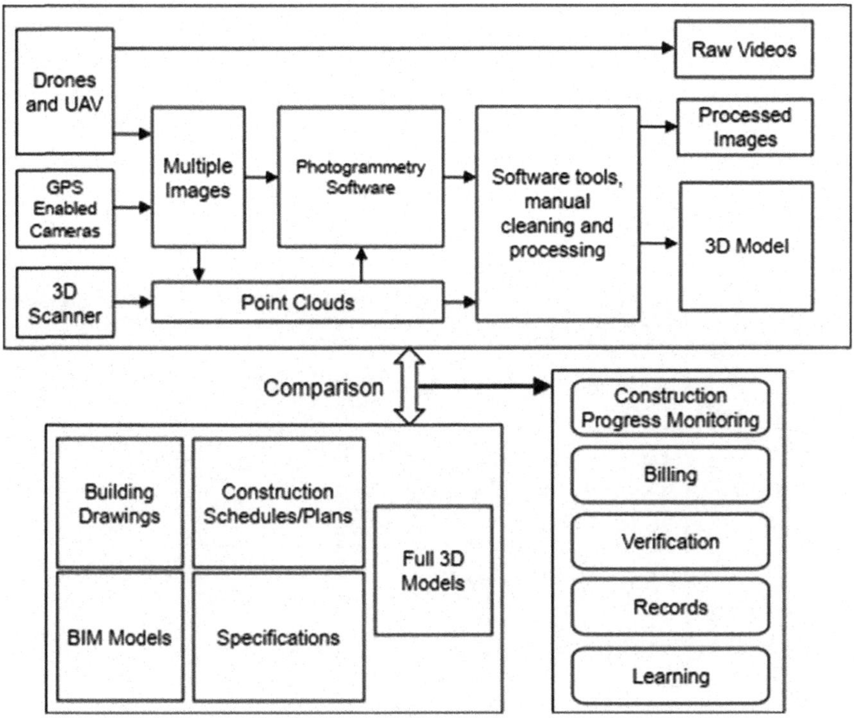

Fig. 4.9 Construction Monitoring/Reporting Concept using Drones and Unmanned Aerial Vehicles (UAV) (Anwar, Izhar & Najam, 2018)

VII. Site Logistics and Planning.

Construction tasks are subjected to human errors even in proper planning, site design, and inventory management. However, the introduction of drone technology in the construction industry helps minimize many human errors. The drone technology makes it possible to capture the entire site used to generate a building design required for proper site planning and logistics. Using drones and Site scan, the project team is availed with the updated and precise site logistics plans. This enables everyone on the project site to be well informed of the locations of gates, truck routes, closures on-site, inclement weather areas, traffic flow, and the list goes on. The project plan can be accessible by anyone on the site, either through a mobile device or computer (Gibson & McManmon, 2018). Figure 4.10 depicts a complete site logistics plan for a project team by drone.

VIII. Thermal Imaging recording.

Drones are equipped with a distinct thermal camera or thermal imaging sensors which can interpret thermal energy (heat) into visible light. Thermal energy, which is invisible to the naked eye, can be illuminated through the infrared camera sensors of the drone. Thermal imaging is required to examine a specific object or location. In construction, drones can generate aerial thermal images to examine cold spots in different parts of buildings. This avails the building professionals the opportunity to identify and rectify: hidden construction defects, improper sealing of windows and roof, water infiltration leaks, moisture intrusion, rot areas, thermal hotspots, leaks areas, delamination, and other areas that may cause damage to the building. Thermal

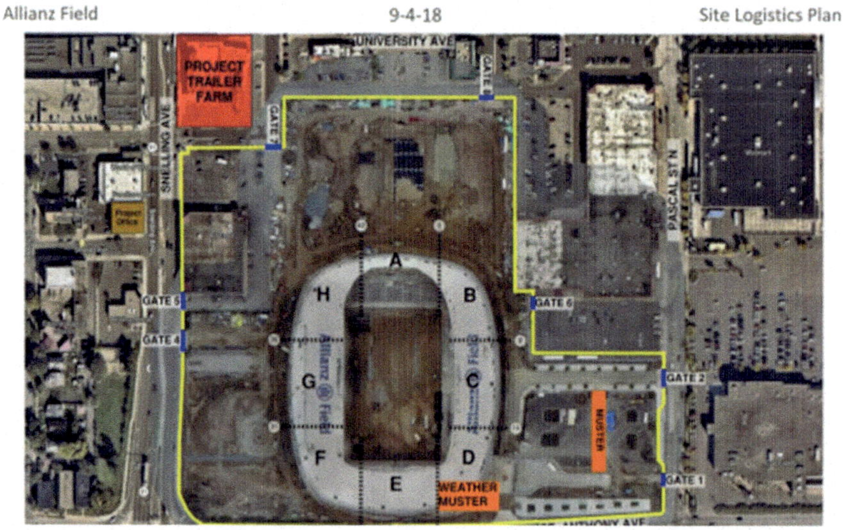

Fig. 4.10 A Case Study of Drones used for Overall Site Logistics Plan for Entire Project Team (Gibson & McManmon, 2018)

Fig. 4.11 Drone Thermal Imaging (Tkac, & Mesaros, 2019)

imaging can also be used to inspect the roof of the building, which gives firefighters insight into heat areas where smoke emanates (Fig. 4.11)

4.3 Case Studies of Drones/Unmanned Aerial Vehicles (UAVs) Technology Application in Construction and Resultant Benefits

4.3.1 Case Study One: 3D Building Construction in Shandong University of Science and Technology and the Henan University of Urban Construction

A case study of the mapping of a 3D building construction in Shandong University of Science and Technology and the Henan University of Urban Construction is examined. In this case study, a Drone/Unmanned Aerial Vehicle with a four-combined camera with a unique design of overlap images was designed and used for these projects (Feifei et al., 2012). The primary use of the drone was for aerial photography and imagery of a 3D building construction with large-scale mapping of 1:1000 and 1:1500. A drone of such capacity was used due to the characteristics and theoretical demonstration that the UAV system for low altitude aerial photogrammetry can be used to construct a 3D building. The most significant advantage of a UAV carrying a four-combined camera is that vertical and inclined images of one building can be captured simultaneously by route planning. For precision, the four-combined camera was regulated individually and collectively. The outdoor calibration field in the Chinese Academy of Surveying and mapping was used for single-camera calibration (Feifei et al., 2012) (Fig. 4.12).

This first 3D building construction experiment in Shandong University of Science and Technology used un-manned airship FKC-1 carrying a four-combined camera for aerial photography. A total of 20 control points of high precision aerial experimental control was filed in the university; the aerial photography meets the aerial photogrammetry accuracy with scale mapping of 1:1000.

The second 3D building construction experiment adopted the fixed-wing UAV carrying a four-combined camera for aerial photography. With intensive control points (36 control points), aerial photography meets aerial photogrammetry accuracy with a scale mapping of 1:500.

4.3.2 Case Study Two: Projects A (91 5-Story Buildings and 5 3-Story Buildings), and Project B (1 26-Story Building)

A case of two residential projects, A and B, are situated in Brazil. Project A consists of residential low-income housing projects with 91 (5-story buildings) and 5 (3-story buildings), while project B consists of a residential high-rise building with 1 (26-story building). Drones were used for construction monitoring, safety inspection, and analysis of the non-compliances based on a safety standard of two projects. Advanced UAV, known as DJI Phantom 3 with Sony EXMOR camera, was used to conduct the monitoring and safety inspection on the construction site. This involved four stages which include targets planning, data collection, data processing, and data analysis. The target planning was carried out to determine the area of interest for monitoring and inspection. At the same time, the data collection with the drone through the visual assets was used to inspect the safety conditions on the construction site. A total of 26 flight tests were performed with an average time of 9 min each. The data collected was fully processed based on the Flight Log Data Form and Safety Checklist by Snapshot Types - Full Version (60 items). A database of safety inspection items and visual assets was created. In total, eight inspections were performed, four for Project A and four for Project B. Due to the large number of visual assets collected, a smaller set of visual assets for safety analysis was selected, which considered many visual assets that could verify all requirements. A total of six to eight visuals assets were selected per inspection. Two analyses were carried out based on the visual assets selected: the first one aimed to evaluate the applicability of the UAV for safety inspection; the second analysis aimed to evaluate the non-compliance with the safety requirement that could be visualized in the first stage, according to the Brazilian Safety Regulations (Santos-de-Melo et al., 2017) (Figs. 4.13 and 4.14).

Fig. 4.12 a The structure of the designed four-combined wide-angle camera. **b** The outdoor calibration field (Feifei et al., 2012). **c** Drone imagery used for mapping 3D building geometric model in Shandong University of Science and Technology (Feifei et al., 2012). **d** Drone imagery used for mapping 3D building geometric model in Henan University of Urban Construction (Feifei et al., 2012)

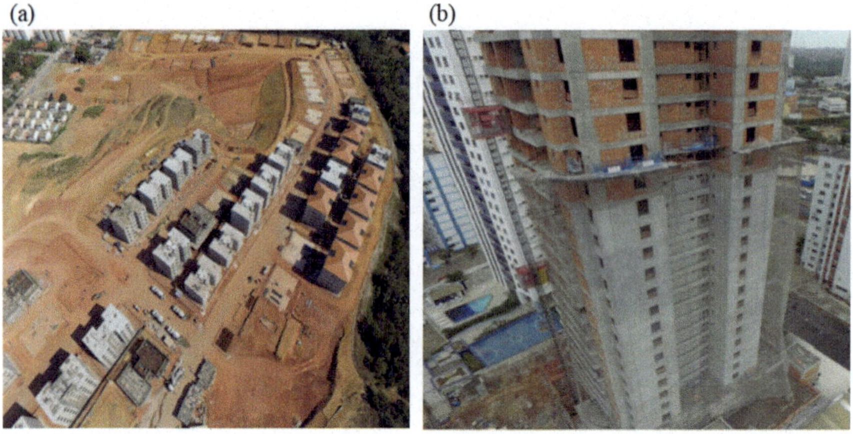

Fig. 4.13 Project A and project B (Santos-de-Melo et al., 2017)

Fig. 4.14 DJI phantom 3 advanced with a tablet Samsung Galaxy Tab E 7 for display (Santos-de-Melo et al., 2017)

4.4 Benefits of Drones/Unmanned Aerial Vehicles Application in Construction

(1) **Facilitate the work of Designers, Engineers, and Contractors**: Drones are becoming an essential component of virtual design and construction (VDC), giving architects and engineers new and efficient ways to visualize and analyze structural requirements from the ground up. The planning, architecture design, and engineering that shape what eventually becomes a building require good data. The raw data captured by the drones help designers create 3D models and maps that can be packaged and sold under various pricing structures.

This becomes yet another tool in the surveyor's arsenal, adding to traditional surveying with modern technology (Danielak, 2018).

(2) **Improve Safety and Minimize Safety Risk**: One of the most profound benefits of drones' technology is enhancing the safety of construction teams on construction sites. Drones are equipped with monitoring devices/components that are used for site surveillance. This technology helps to identify hazardous areas on time and ensure that equipment and materials are appropriately stationed where required on the construction site. As a result, tasks needed by personnel to ensure safety on sites are taking over by drones, and all information needed by the construction crew is easily accessible without risking any life. Hence, minimizing the vulnerability of the workers to accidents, injury, and other risks.

(3) **Enhance Speedy Progress Monitoring on Construction Site**: A major task of project managers is to conduct periodic inspections on construction sites to ensure the quality of projects is enhanced by implementing drone technology. Drones are designed with a camera that can capture all the angles of a construction site within a short time. The visual data collected avails the project managers, teams, and project owners with the updated situation of the ongoing projects, which facilitates the project progress. Also, the drone maps that give a bird view of the construction sites lessen the time spent on inspection, thereby speeding up the monitoring process. Consequently, improvement in progress monitoring ultimately improves the project delivery.

(4) **Saves Cost of Project Execution**: Generally, construction works are time-consuming. However, the introduction of modern technology such as drones represents a cost-saving tool in the construction industry. This is achieved with the operation of drones for surveillance and aided inspection on the construction sites. This helps to monitor and identify suspicious activities and errors that might lead to delays and ultimately result in cost overruns. Also, with regards to risk management, the cost of drone technology insurance is lower than the on-site workers.

(5) **Saves Time**: The introduction of drone technology in the construction industry for tracking, mapping, automation, and surveillance has enhanced speed in construction activities. Data collected manually on the time-consuming construction sites can be generated speedily using drones. Similarly, the small size of drones makes it possible to fly and maneuver swiftly to capture the entire project site and send back photographs and footage of all the information required for effective project execution. This saves a lot of time that would have required the construction workers and professionals to achieve.

(6) **Facilitate Communication**: Communication across the construction crew must keep tabs on the construction progress; this has been made easier through drone technology. Drones with a cloud-based facility and mounted make it possible to share annotated maps and provide video footage that facilitates communication and collaboration among the project teams. This visual data deduced from the drone allows all team members and external stakeholders

to assess the construction sites, thereby enhancing communication. A clear display of the updated site situation represents an important communication tool.

(7) **Enhanced Decision-Making Process**: Decision-making process is vital in achieving a quality project within a stipulated schedule and cost. Drones give a bird's eye view of a construction site that promotes accurate decision-making on real-time information instead of narratives. The drone technology enhances the significance of data in the construction process. Drone technology improves decision-making through quick data collection, mapping, and analysis.

Furthermore, Unmanned Aerial Vehicles avails project managers updated visual data of daily activities such as topography and aerial, surveying, analysis of stockpiles, measurement of excavation depths, materials movement, safety and security, and monitoring of construction sites progress. The information collected, which can be reviewed, benchmarked, archived, and compared with other information, is a valuable tool for better decision making. Also, the decision-making process can be streamlined by referencing a drone map which allows for changes to be accommodated anytime.

4.5 Risks Associated with Drone Technology

Rapid use in drone technology, especially for commercial purposes such as construction, presents several benefits and, at the same time, poses some risks. These risks emanate from technical, legal issues and uncertainties. Some of these risks are discussed below.

(a) **Drones Registration and Regulations**: Different countries have established rules and regulations regarding licensing and use of drones. Some of these regulations specify adherence in weight, height, the purpose of drone use, certification, and obligations of drone operators. When these regulations are not strictly adhering to, it poses the risk of prosecution by the Civil Aviation Authority (CAA). This might result in large fines, withdrawal of license, and even imprisonment. This can be corroborated with Yaacoub et al. (2020), which states that violators of drone regulations could be penalized with unlimited fines or a jail term of five years. Consequently, the ongoing construction projects are obstructed, leading to cost overruns and delays in the projects' delivery.

(b) **Data Privacy**: In construction, a drone is considered a data capturing technology that is used to collect real-time data of construction projects and the activities on the construction site. As a result, the technology is susceptible to data theft, trespass, invasion of privacy, and intellectual property. Some of the risks include data theft or cybersecurity, resulting in drone data and imagery records being stolen or compromised by a competitor. Drones operators' preference for using smartphones and devices for operation makes data collected highly susceptible to hacking. Also, data privacy risk can occur when a drone operator secretly captures personal images, including workers, visitors, and third parties, recordings, or entering another personal property without authorization. The recognition of some of these risks has informed Federal Aviation Authority (FAA) (2018) in formulation rules and regulations guiding drones.

(c) **Accident**: Another potential risk of drone application in construction is drone accidents. Drone accidents on construction projects have been a prominent occurrence in the industry. Drones accident occurs when there is a collision with another object, unfavorable weather condition, hijacking or attack by external parties or drone crashing. Consequently, this can lead to property damage, personal injury, and loss of life. The aftermath of the drone accident ultimately impacts negatively on the delivery of the projects under execution. It is important for construction firms to secure proper insurance cover considering the risk of drone accidents.

To manage the risks involved in drone use, the construction industry, firms, and participants should get acquitted with rules and regulations on the use of drones specified by the Federal Aviation Administration (2018). Also, there is a need for project managers to constantly enlighten drones operators on the regulations and develop a procedure for managing the risks involved in drone usage. An individual construction company can build its regulations and procedure to guide use of drones for project execution. These include drone planning and operations, conditions and processes for drone use, cybersecurity, intellectual property, establish a data management team, qualifications, training, and compensation of drone operators. It's also essential to get insurance cover and secure the service of a well-grounded lawyer who specializes in the drone industry.

4.6 Skills Required for Drone Operation in the Construction Industry

Drones application requires specific skills and knowledge to be considered as a good drone operator. This is vital across all sectors where drones are used as safe drone flights trained and licensed pilots. In today's world, drone technology has become an indispensable device, and securing the required skills becomes a vital tool for any industry to gain a competitive advantage. Also, considering the ecosystem, knowledgeable and skilled drone personnel are necessary for various roles such as maintenance, innovation, and development of products and solutions. In general terms, a drone operator is expected to be skilled in video and camera settings, photography, filming technique, manual flight. However, modern Unmannered Aerial Systems (UAS) are becoming more autonomous, requiring higher skills than general skills. Considering the development in the use of drone technology, the following are the core skills for drone application.

(1) **Communication Skills**: Drone operations require solid and effective communication as it relies on transferring information and data. There are several communications in drone technology. Yaacoub et al. (2020) classified drone communication into Drone- to-Ground-Station (D2GS), Drone-to-Satellite (D2S), Drone-to-Drone (D2D), and Drone-to-Network (D2N). Communication is vital in the construction industry, and drones are currently primarily used for communicating with all parties, onsite and offsite, involved in project execution. Consequently, drones operators require specific communication skills for effective operation.

(2) **Programming Skills**: The introduction and application of new technologies that demand computer programs have made programming skills crucial. A drone operator requires programming skills to manage, process, and analyze correctly the data collected by the drone. Advance Unmannered Aerial Vehicle Systems are becoming more autonomous with several systems of autopilot modes. This demands a high level of programming skills to successfully activate the flight assistance tools of drones such as homing, levelling, automatic takeoff, and landing methods. Construction dynamics would always give rise to reprogramming of worksite drones if the drones worked autonomously. Programing skills in drone technology could be outsourced to experts in the absence of employed professionals or funds to finance a permanent drone expert.

(3) **Data Analytics Skills**: A drone or unmannered aerial system is a -data-driven technology used to capture, store, and transmit data. The advancement in the application of the unmannered aerial vehicles system has expanded beyond just capturing videos but used for wider applications that require the collection of data for surveys. These technologies are now designed to produced large datasets, which demands proficiency in data analytics. For the application of drone technology in the construction industry, data analytics skills are required to extract valuable information from the drone, analyse and deliver data accessibility to the project teams to make informed decisions. This skill is also to address critical problems that might arise regarding data generated by the drone.

(4) **Machine Learning Skills**: Drones can be categorized as one of the new automated technology of the fourth industrial revolution. Machine learning is essential to automate the analysis of aerial data. With machine learning proficiency, it is possible to extract information from data, make precise measurements, detects objects and transform extensive data into consumable reports. Machine learning has proffered solutions to some of the challenges faced in an unmannered aerial system for autonomous flight in recent times. Proficiency in machine learning can help identify and predict data patterns alongside software programming for autonomous flight.

4.7 Summary

The drone technology represents one of the fourth industrial revolution technologies (4IR) that has gained wide application and experienced rapid technological development. These technologies, also referred to as Unmanned Aerial Vehicles (UAVs), have become indispensable and digitally transformed the construction industry. In recent times the capabilities of drone technology have increased by the inclusion and pairing with several vehicles such as sensors, GPS navigation units, laser scanners, and high-resolution cameras for commercialization around the world. There are several variations of drone technology, and these can be classified in terms of operating weight/mass and the purpose of drone flight. In the construction industry, drone technology has rapidly gained much acceptance and become an essential tool applied in all the stages of the construction process. The use of drones has presented several benefits to construction projects. These include aerial photography resulting in precision in reporting, improved safety, better inspection, speedy progress monitoring on construction sites, cost-saving project execution, and enhanced decision-making process. The increasing use of drone technology has brought about the formulation of stringent policies and restrictions by a legal framework to manage the risks of drone use, such as data privacy, accidents, hijacking and coalition that caused damage from unauthorized pilots. The advancement and modification of the drone or Unmanned Aerial System (UAS), characterized by automation and robotics, demands specific critical skills other than universal skills. Skills such as machine learning, programming, data analytics and communication will be required by drone operators to effectively utilize the technology to enhanced sustainability in the construction industry.

References

Accenture (2019). A business approach for the use of drones in the Engineering & Construction industries. Retrieved August 1, 2019, from https://www.accenture.com/_acnmedia/pdf-24/accentu redrones-construction-service.pd

Albeaino, R. G., Gheisari, M., & Franz, B. W. (2019). A systematic review of un-manned aerial vehicle application areas and technologies in the AEC domain. *Journal of Information Technology in Construction, 24*, 381–405.

Anwar, N., Izhar, M. A & Najam, F. (2018). Construction monitoring and reporting using drones and unmanned aerial vehicles (UAVs). In: *The Tenth International Conference on Construction in the 21st Century (CITC-10). Colombo, Sri Lanka.*

Ayemba, D. (2018). Utilizing drone technology in construction. Retrieved March 18, 2020, from https://constructionreviewonline.com/2018/03/drones-in-construction/ on 18/03/2020.

Danielak, M. (2018). The benefits of employing drones in construction. Retrieved from https://www.constructiondive.com/news/the-benefits-of-employing-drones-in-construction/516713/

Devers, W. (2019). *A case study on the use of drones on heavy civil construction projects.* Retrieved from Digital commons: https://digitalcommons.calpoly.edu/cgi/viewcontent.cgi?article=1324&context=cmsp

DeYoung, R. E. (2018). An Industrial Overview for The Standardization of Drones in Construction. Unpublished Masters thesis, Everglades University, Florida.

Federal Aviation Administration. (2018). FAA Aerospace Forecast Fiscal Years 2018–2038. Retrieved from https://www.faa.gov/data_research/aviation/aerospace_forecasts/media/FY2018-38_FAA_Aerospace_Forecast.pdf

Feifei, X., Lin, Z., Gui, D., & Lin, H. (2012). Study on construction of 3D building based on UAV images. *International Archives of the Photogrammetry, Remote Sensing and Spatial Information Sciences, XXXIX-B1*, 469–473.

Ford, J. (2018). *The History of Drones (Drone History Timeline from 1849 to 2019.* Retrieved from Dronethusiast: https://www.dronethusiast.com/history-of-drones/

Fox, S. J., (2017). The rise of the drones: Framework and governance—why risk it! *Journal of Air Law and Commerce, 82*(4), 683 – 715. Retrieved from https://scholar.smu.edu/jalc/vol82/iss4/2

Gibson, A., & McManmon, J. T. (2018). *A case study on drones in construction: Easier than you may think.* Retrieved from Autodesk: https://www.autodesk.com/autodesk-university/class/Case-Study-Drones-Construction-Easier-You-May-Think-2018#downloads.

Giordan, D., Adams, M. S., Aicardi, I., Alicandro, M., Paolo, A., Marco, B., et al. (2020). The use of unmanned aerial vehicles (UAVs) for engineering geology applications. *Bulletin of Engineering Geology and the Environment, 79*, 3437–3481. Retrieved from https://doi.org/10.1007/s10064-020-01766-2.

Grayson, W. (2016). How drones and UAVs are already affecting construction job sites. Retrieved August 12, 2016, from http://www.equipmentworld.com/drones/

Hassanalian, M., & Abdelkefi, A. (2017). Classifications, applications, and design challenges of drones: A review. *Progress in Aerospace Sciences, 91*, 99–131. Retrieved from https://www.researchgate.net/profile/Mostafa_Hassanalian/publication/316673697_Classifications_applications_and_design_challenges_of_drones_A_review/links/59ca8917aca272bb05078e84/Classifications-applications-and-design-challenges-of-drones-A-review.pdf.

Higgins (2017). "Beginner's Guide to Using Drones for Construction Management." Connect & Construct. Retrieved from www.connect.bim360.autodesk.com/drones-for-construction-management.

Kazi, M. H., Newaz, S., & Ahsan, M. S. (2018). Design and development of an aircraft type portable drone for surveillance and disaster management. *International Journal of Intelligent Unmanned Systems*, 1–16. Retrieved from https://doi.org/10.1108/IJIUS-02-2018-0004.

Liew, C. F., DeLatte, D., Takeishi, N., & Yairi, T. (2017). Recent developments in aerial robotics: A survey and prototypes overview. Retrieved from https://arxiv.org/pdf/1711.10085.pdf?utm_content=buffer9a5b8&utm_medium=social&utm_source=twitter.com&utm_campaign=buffer.

Lin, A. (2018). *Use of drone technology on commercial construction projects.* Retrieved from Digitalcommons: https://digitalcommons.calpoly.edu/cgi/viewcontent.cgi?article=1170&context=cmsp.

Ostojić, G., Stankovski, S., Tejić, B., Đukić, N., & Tegeltija, S. (2015). Design, control and application of quadcopter. *International Journal of Industrial Engineering and Management, 6*(1), 43–48. Retrieved from https://www.researchgate.net/publication/283110601.

Papa, U., Ponte, S., & Del Core, G. (2017). Conceptual design of a small hybrid unmanned aircraft system. *Journal of Advanced Transportation, 9834247*, 1–10. https://doi.org/10.1155/2017/9834247.

Sanson, J. S. (2019). Drone use in the construction industry leads to integration into the current civil and construction engineering technology curriculum. In: *Proceedings of the 2019 Conference for Industry and Education Collaboration.* Retrieved from: https://peer.asee.org/drone-use-in-the-construction-industry-leads-to-integration-into-the-current-civil-and-construction-engineering-technology-curriculum.pdf.

Santos-de-Melo, R. R., Costa, D. B., Álvares, J. S., & Irizarry, J. (2017). Applicability of unmanned aerial system (UAS) for safety inspection on construction sites. *Safety Science, 98*, 174–185. Retrieved from https://doi.org/10.1016/j.ssci.2017.06.008

Tatale, O., Anekar, N., Phatak, S., & Sarkale, S. (2018). Quadcopter: Design, c\onstruction and testing. *International Journal for Research in Engineering Application & Management, 04*(Special Issue AMET-2018), 1–7. https://doi.org/10.18231/2454-9150.2018.1482

Tatum, M. C., & Liu, J (2017). Unmanned aircraft system applications in construction. In: *Creative Construction Conference 2017, CCC 2017, 19–22 June 2017, Primosten, Croatia Procedia Engineering* (Vol.196, pp. 167–175)

Tkac, M., & Mesaros, P. (2019). Utilizing drone technology in civil engineering. *Journal of Civil Engineering, 14*(1), 27–37. https://doi.org/10.1515/sspjce-2019-0003

Yaacoub, J. P., Noura, H., Salman, O., & Chehab, A. (2020). Security analysis of drones systems: Attacks, limitations, and recommendations. *Internet of Things, Elsevier Public Health Emergency Collection, 11*, 100218. https://doi.org/10.1016/j.iot.2020.100218.

Young, J. (2019). The use of drones in construction Inspection and general building. Retrieved August 4, 2019, from http://www.droneguru.net/the-use-of-drones-in construction inspection-and-general building/

Zitzman, L. (2018). Drones in construction: *How They're Transforming the Industry*. Retrieved August 2, 2019, from https://www.bigrentz.com/blog/drones-construction

Zhou, S., & Gheisari, M. (2018). Unmanned aerial system applications in construction: A systematic review. Construction Innovation, CI-02–2018–0010. https://doi.org/10.1108/CI-02-2018-0010

Chapter 5
Prefabrication

5.1 Introduction

Prefabrication is a building technique that has made it possible for incomprehensible building structures to be built within a little time. Although this building method is considered one of construction 4.0 technologies, it has earlier existed during the post-war era but later went into extinction. However, technological innovation has created more attention/ interest in prefabrication in the construction industry. Prefabrication existed during the ancient period of burnt clay bricks of the Mesopotamian civilization in the Roman forts, when the conqueror brought prefabrication sections of defences for invading England to post-world war II housing (Redshift, 2019). During the period, a panelized wood house, the first known prefabrication, was shipped from England to Massachusetts to housing a fishing fleet. Majorly, the historical background of prefabrication mould-making in construction is associated with the Greek and Roman. Hence, this building system has been recognized as one of the alternative solutions to changing the speed of conventional construction methods at a fast rate.

Also, one of the notable events in the historical foundation of prefabrication is the rescue mission, which helps provide a solution for the accommodation for families after world war in 1946 when buildings were destroyed (Hearn 2018). In recent times, prefabrication/modular construction is increasingly becoming attractive and gaining much recognition due to its bail-out for housing shortages experienced in some parts of the world. The recent innovation in this building technique has led to the speedy realization of good quality projects such as skyscrapers, bridges, offices, hospitals etc.

The term prefabrication has been described using several words: off-site manufacturing, modular construction, industrialized building, etc. The most common term used interchangeably is prefabrication with modular construction. This difference in these terms shows that modular construction is just a classification according to the structural configuration, where almost 75% of construction is done at a factory site. In the construction industry, the term prefabrication describes building components manufactured in an industrialized setting and then transported to construction sites

O. Adepoju et al., *Re-skilling Human Resources for Construction 4.0*, Springer Tracts in Civil Engineering, https://doi.org/10.1007/978-3-030-85973-2_5

to be assembled into a building structure. It can be classified as a construction or manufacturing process with the end-product of a preassembled component prepared at a dedicated site with different construction materials. This end-product includes bricks, slab pairs, columns, window shades, wall panels, and a complete construction component (Muhammad et al., 2017). Prefabrication as a manufacturing process takes place at a specialized facility, in which the various building materials are joined to form a part of the final installation (Zhong-Lei et al., 2019). The fabrication of different building components at an offsite location separate from the construction site is carried out to improve control, reduce labour operation on-site, and better utilise resources—all in the interest of speed and profit.

5.1.1 Classification of Prefabrication

There are several classifications of prefabrication adopted by different authors based on the system of operation. Adapting from the work of Alireza et al., 2015; Fred et al., 2016, prefabricated systems can be classified according to structural configuration, materials and components.

1. **Structural Configuration:** Prefabrication under structural configuration can be classified into panel system, frame system, box/module/cell system and hybrid System.

   (i) Panelised systems of prefabrication are units produced in a factory and assembled on-site. This system boosts the speed and convenience of conveying walls to a site (Figs. 5.1 and 5.2). An example of a panel system is the 30-story hotel near Dongting Lake in the Hunan Province of China, called T30 that was built in 15 days (Lauren, 2015).

   (ii) **Frame system:** Frame system provides the possibility of erecting buildings of any geometrical configurations and the different storeys of the same building (Figs. 5.3 and 5.4). Roof and walls systems are joint to the frame by bolts that exclude welding construction parameters that meet every client's requirements regarding volume and space. (Anuradha & Dhawade, 2015).

   (iii) **Modula system:** The modular system, also known as the box or cell system, is a three-dimensional construction modular unit mass-produced completely with services and finishes equipped in a factory setting before being transported to the construction site (Tharaka et al., 2016). This prefabrication system is made up of 3D box-like modules that can serve as both interior and exterior designs. These modules can be easily assembled and quickly erected. An example of a modular concept is the mobile home which can be easily moved from one place to another (Figs. 5.5, 5.6 and 5.7). A typical case study is the Mini Sky City which is a 67 storey skyscraper apartment constructed in 19 working days (Fred et al., 2016).

Fig. 5.1 An exterior wall panel (John, 2020)

Fig. 5.2 Panels assemble on site (Anuradha & Dhawade, 2015)

(2) **Materials:** Prefabrication is classified based on the materials used as the major build-up of the building structure. Prefabrication units are made up of some basic construction materials, including concrete, steel, and wood. Recently, glass fibre reinforced polymers were included as one of the materials for prefabrication (Alireza et al., 2015; Martin, 2019).

(i) **Concrete:** Concrete modular buildings are not very popular. In high rise prefabricated buildings, concrete is mainly used for the panels of partially prefabricated buildings or as a core to be combined with the modular units (Martin, 2019). Some of the advantages of concrete

Fig. 5.3 Frame system (Anuradha & Dhawade, 2015)

prefabrication materials include high quality, durability, strength, and withstanding/endurance of heat, thereby reducing the energy consumption of the building. The disadvantages of concrete materials include waste, which leads to high transportation costs and sustainability due to difficulty in recycling or reuse.

(ii) **Wood:** In the past, wood has been the most used structural material in prefabrication, especially with modular methods. This is because modular construction has been for a long time associated with the family home detached houses. The wood material is also often used due to its low price, good performance, and high solidity value recognized by designers. Other advantages of wood include renewable and energy-efficient, which is associated with low-rise buildings and ease of transportation due to its lighter weight than other materials. The disadvantages of wood are the limitation in height requiring a maximum amount of floor connection and the inability to withstand loads during placement and transportation. Hence, installing temporary bracing to cushion/resist loads (Hopkins, 2014; Martin, 2019).

(iii) **Steel:** Steel is also used in modular buildings for a structural system. Steel is a metal structure fabricated that can be used to construct a steel building. It can be used for interior support and exterior covering. The steel material for prefabrication has the advantages of: durability,

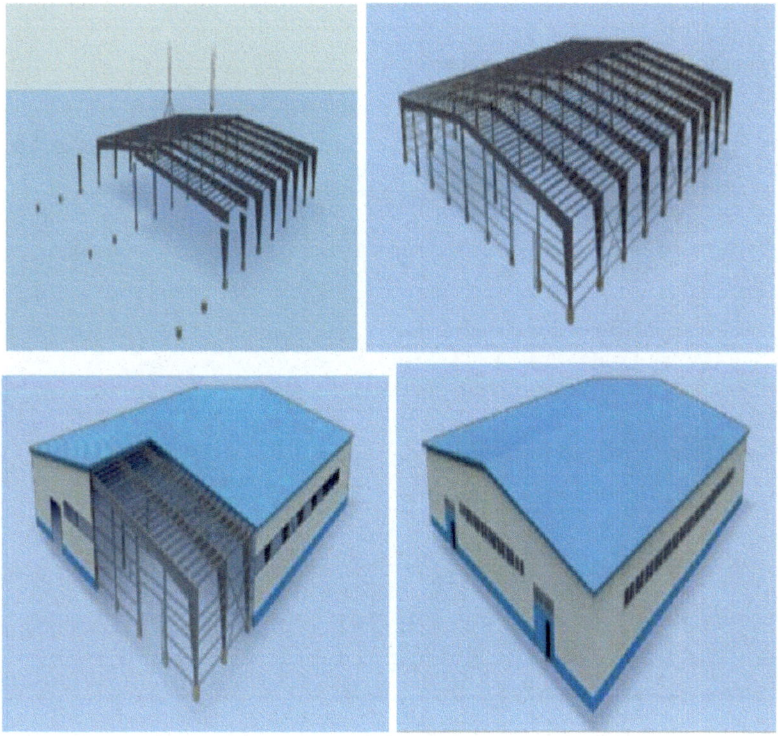

Fig. 5.4 Frame assembly process (Anuradha & Dhawade, 2015)

firmness, lightweight, recyclability, and ability to resist lateral force. These advantages reduce the amount of work that ought to be done on-site, thereby saving cost and time. The rigid and robust structure allows larger opening spans, reduced need for lateral bracing and more design flexibility (Martin, 2019).

(iv) **Fiberglass:** Fibreglass is also referred to as fibre reinforced plastic using glass fibre, which can be rearranged, flattened into a sheet or woven into a fabric and used to protect other materials such as concrete, steel and wood. In prefabrication, fibreglass is often used for decoration and beautification. The aesthetic formability of fibreglass has made it the most preferred interior and exterior design structure and coating material (Polser, n.d). Also, in architecture, it can be used for insulation which comes in roof and insulation panels. The advantages of fibreglass include flexibility (it can be designed into several shapes), low cost, durability, high impact resistance (to water, weathers, corrosion etc.) and low thermal conductivity serving as an excellent insulating material.

Fig. 5.5 A Module or Unit (Fred et al., 2016)

(a) (b)

Fig. 5.6 An image from the modular building: **a** After being built and occupied; **b** During its on-site assembly (Navaratnam et al., 2019)

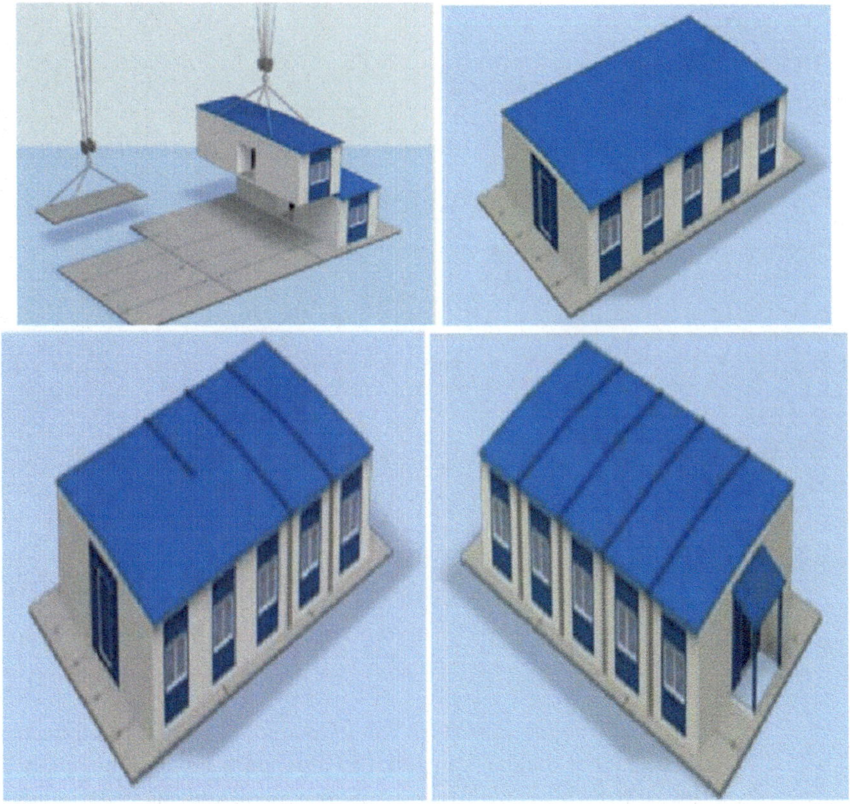

Fig. 5.7 Construction building process of modular components (Anuradha & Dhawade, 2015)

(3) **Components:** Prefabrication components can be classified into vertical and horizontal.

(i) **Slabs:** The prefabricated slab is one of the parts of contemporary buildings which comprise a horizontal flat surface made of cast concrete. Slabs are often used to construct ceilings cast in a factory and transported to the site through trucks. It can be categorised as a suspended or ground bearing, resting on a foundation. Examples of slabs used in construction include flat slab, precast slab, composite slab, hollow core slab and one-way joist slab (Fig. 5.8).

(ii) **Beams:** The beam is a horizontal structural component produced with high strength concrete that can withstand load bending moments and shear forces that transfer load from slabs to the columns. It gives support to the slab by resting on the column. Beams are designed to fit to the structural requirement and according to the sizes and shapes of the columns provided. The typical depth of the beam is 16 to 40 inches, and the typical width provided is 12 to 24 inches (Fig. 5.9). Prefab

Fig. 5.8 Prefab slab (Chinmay, 2019)

beams and columns provide flexibility in design, shape and application and are highly durable compared to alternative building materials. It offers a clean, finished surface for the structural aesthetic of the building (Chinmay, 2019).

(iii) **Column:** The column is considered the most important structural member as it helps to transfer huge loads from beams to the foundation. It is a vertical supporting pillar that can be in the form of a single or double-storey height. Columns are initially cast in hollow sections with an arrangement of dowelled connections. Prefab columns can be erected five times faster than cast-in-situ columns (Fig. 5.10). It can achieve a high quality and uniform finish. Erection of a column is a silent process that can take place at any time. (Chinmay, 2019).

(iv) **Shear Wall:** A Shear Wall is a vertical structural element composed of braced panels (also known as shear panels), which is used to counter the effect of lateral load acting on a structure. Shear wall is usually designed to resist wind in-plane and seismic loads. The applied load is generally transferred to the wall by a diaphragm or collector (Fig. 5.11). Typically, materials such as wood and plywood are used in shear walls. However, advancements in technology have introduced more options for sharing walls in prefabrication, such as steel and concrete. Hence, creating shear assemblies into narrow walls that fall at either side of an opening and can be cast according to the design requirements. (Chinmay, 2019).

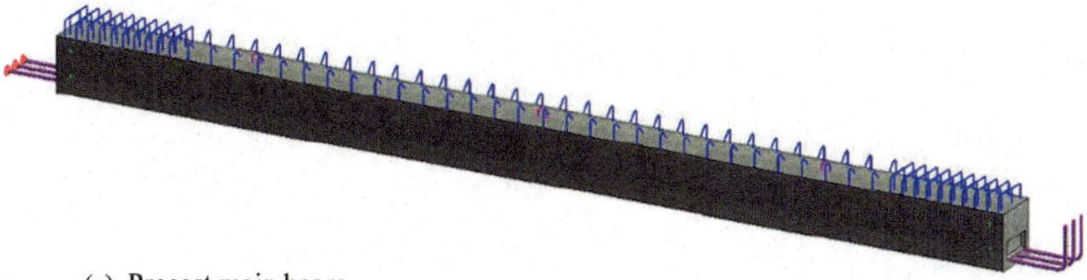

(a) Precast main beam

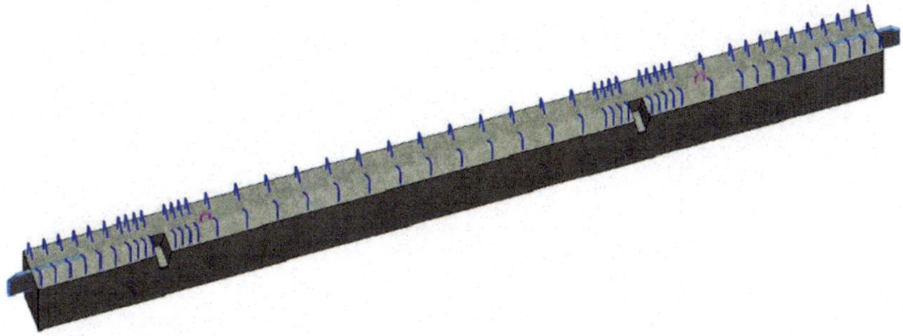

(b) Precast secondary beam

(c) Connecting detailing between the prefabricated main beam and secondary beam

Fig. 5.9 Prefabricated Beam (**a**), (**b**), (**c**), (Lida et al. 2018)

Fig. 5.10 Precast column (Chinmay, 2019)

5.2 Application of Prefabrication in Construction

In time past, prefabrication techniques are applied to housing facilities, but in recent times, technological evolutions have brought about a paradigm shift to other forms of construction. These include bridges, highways, stadiums, shopping centres, health-care facilities, skyscrapers, schools, churches, ramps and more. The application of prefabrication can be grouped as follows:

- Residential and Commercial Housing
- Building, Rehabilitation and Reconstruction of Bridges
- Stadiums and Grandstands.

Fig. 5.11 Prefab share wall (Chinmay, 2019)

5.2.1 Residential and Commercial Housing

From history, housing had been a significant challenge. The advent of prefabrication paved the way for housing accommodation for communities and apartments to host soldiers who returned from war. For many years, prefabrication has been applied to residential buildings, but recently it has extended to commercial buildings. Prefabricated residential and commercial buildings are built by industrial method, which entails assembling all building components prefabricated in the factory and later transported to the construction sites. In recent times, several countries worldwide are now adopting prefabrication to construct their residential and commercial outfits. Architects and designers across countries have adopted prefabrication construction techniques for sustainable and low-cost housing. Prefabricated residential housing are houses designed to provide accommodation for people or families either temporarily or permanently. These include private apartments, hotels, hospitals etc. Figure 5.12 shows the interior and exterior of permanent modular school buildings, an initiative of the Victorian School Building Authority (VSBA) of Australia to

Fig. 5.12 Permanent modular school buildings (Navaratnam et al., 2019)

replace old schools' buildings. On the other hand, commercial housing includes shopping malls, kiosks, showrooms, offices etc. Figure 5.13 depicts a prefabricated office building with a complete manufactured component at an off-site, quality-controlled factory.

5.2.2 Reconstruction of Bridges and Rehabilitation

Across the world, road traffic coupled with an ageing road network that needs repair is vastly becoming a crucial issue that demands critical attention to minimize traffic and ensure free flow movement on the highways. Consequently, this requires speedy re-construction or renovation of the roads and bridges to avoid road closures, leading to other complications. These challenges have been resolved through the adoption of prefabrication. Modern technology in construction has to offer new techniques of prefabricated bridge such as Prefabricated Bridge Element System (PBES), Prefabricated Bridge System (PBS) and Prefabricated Precast Bridge System (PPBS). This prefabrication mode consists of variations and a blend of different precast elements made of concrete or steel or combining the two. In recent times, Government agencies of several countries across the world, especially the developed ones, have resolved to the application of prefabrication to accomplish quality road and bridge construction projects within a short period. Also, the adoption of prefabrication offers several advantages over on-site casts, such as cost reduction due to off-site manufacturing reduces traffic disruption. It improves safety by minimizing workers' and environmental risk encounter during construction. The adoption of prefabrication for road and bridge construction makes it possible to speed up work two to three times against

Fig. 5.13 Modular building for offices (Navaratnam et al., 2019)

the traditional construction approach (Radan, 2017). Figure 5.14 depicts modular bridge construction.

5.2.3 Stadiums and Grandstands

The application of prefabrication for the construction of stadiums and grandstands has also gained recognition in recent times. This is majorly applied with the use of precast concrete for building stadiums. Architects, contractors and clients have discovered the use of prefabrication as an exceptional application for outdoors and indoors facilities for precast stadiums. Although not all parts of the stadiums are fully prefabricated, the application of its construction techniques to building parts had made a glaring impact in the construction of stadiums. The current requirement of stadium designs regarding vibration control, durability and fast construction speed makes prefabrication techniques a perfect application. Prefabrication stadiums

Fig. 5.14 Kintaikyo Bridge Reconstruction Project. **a** New arch ribs. **b** Falseworks for arch construction. **c** After replacement (Mohiuddin, 2015)

constructed are entirely different from conventional stadiums based on the advantages it offers. It is more durable, allow shorter timeframes, lower programming costs without any compromises in terms of flexibility, comfort, safety and has light tolerance control, produce fast erections and champion quick stadium completion.

5.3 Case Studies of Prefabrication Application in Construction Projects and the Resultant Benefits

5.3.1 Case Study One: Fort Sam Houston Medical Education and Training Complex

Modular and onsite construction was adopted in a Medical Education and Training Complex construction project in San Antonio, Texas, Fort Sam Houston. This construction, which is expected to house 2400 medical trainees, was accomplished with three construction techniques. The construction techniques are factory-built dormitory modules that utilized a wooden shear wall design, steel frame construction, and the moment frame construction that uses rigid-welded connections at the intersection of a column and beam to resist lateral loads. The use of Permanent Modular Construction (PMC) offered an effective solution to speed up the project's construction without compromising the quality, ensuring the safety and durability of the building. Prefabrication helped to realize the desired requirements of the clients, thereby creating a LEED-certified building with a total square footage of one million sqft. This is the world's largest military medical training facility and one of the world's largest permanent modular construction projects in the world. Thus, this project achieved a state-of-the-art structure built at an affordable price in a relatively short period (Fig. 5.15). Therefore, a quality, safe, sustainable, and superior result is achievable by effectively utilizing prefabrication and modularization for construction projects (Ankit, 2016).

5.3.2 Case Study Two: One9 Modular Building in Melbourne Australia

One 9 is a ten-storey modular building consisting of 34 apartments is located in Melbourne central district in Moonee Ponds, Australia. This construction was designed by Amnon Weber architecture firm and constructed by Vaughan Constructions using a prefabricated building system. The modular ground floors were assembled on-site, and the manufactured apartments were erected using 36 unitized building modules in five days. The building was situated on a land area of 277 square metres, which is considerably small. Due to the heavy site environment congestion and fear of disturbance considering the shopping mall located close by, prefabrication technology has been proven as the best option (Fig. 5.16). This has helped save costs that would have otherwise been incurred due to challenges encountered in construction in a heavily congested suburb with a difficult set of constraints. Adopting prefabrication techniques didn't leave amenities off the books, as the building had secured a six-star rating from Green Star Australia for its many features that endorse sustainable construction (Tharaka et al., 2017).

Fig. 5.15 Fort Sam Houston medical education and training complex in San Antonio, Texas (Ankit, 2016)

5.3.3 Case Study Three: La Trobe Tower in Melbourne, Australia

La Trobe is the world's tallest modular building with 43 storey construction located in central Melbourne CBD that consists of 206 residential apartments. This project was developed by Long river Group, designed by Lothelowman architects and executed by Hickory. The projecting tower, which is 133 m high, was completed and delivered in 19 months. Thus, it shows a 60% faster than convention construction. The project used structural building elements patented by hickory, sync bathroom pods and structural construction elements such as shotcrete, precast and load-bearing concrete wall elements. This project has succeeded in carrying out entirely night-time construction to ensure minimal disturbance to the rest of the city. More importantly, it has served as a productive solution for post-disaster relief operations (Fig. 5.17). This method of prefabrication presented more excellent safety on-site, reduced the number of crane lifts to accelerate the construction process and delivered a quality project on time (Tharaka et al., 2017).

Fig. 5.16 One9 Nine Storey Prefabrication apartment Tower (Tharaka et al., 2017)

5.4 Benefits of Prefabrication Application in Construction

(1) **Saves time:** A prefabricated system provides a substantial decrease in time. Compared to conventional construction methods, prefabrication brings about a 40% decrease in the time required to complete an onsite construction project. A major contributing factor to this benefit is the already prefabricated building components that eliminate delays of scaffolding and shuttering on the construction site. The prefabrication method of the building also gives contractors the chance to take up multiple projects due to preparing the construction site alongside the fabricated modules.

(2) **Reduce cost:** Any construction work aims to get a quality structure in place at the least price. Prefabrication construction reduces the cost and time of the on-site construction phase significantly compared to conventional construction (Navaratnam et al., 2019). Prefabrication involves bulk production and

Fig. 5.17 43 level prefabrication la trobe tower in Melbourne, Australia **a** Design tower form **b** View of the tower (The State of Victoria Department of Environment, Land, Water and Planning, 2015)

purchase, which helps achieve cost reduction in the procurement and transportation of construction modules. Also, prefabrication offers the computerization of the construction process. This permits a high degree of customization at an affordable cost. The fact that time is saved due to ready-made components for building speed results in a low labour cost per project. In sum, prefabrication is adopted where labour costs, power, materials, space and overheads are lower, thereby achieving economy in cost.

(3) **Reduce waste:** Prefabricated construction systems reduce wastage of materials than in site-built construction. This is possible as most construction operations occur in the factory, thereby minimizing potential errors during the construction process. Adopting prefabrication reduces theft of materials and damage due to vandalization or environmental hazards. Prefabricated building modules can be disassembled, relocated, or retrofitted and refurbished to be used in other projects, decreasing waste due to excesses and surplus. (Navaratnam et al., 2019). Also, wastage of materials can also be minimized as prefabrication avails the opportunity to recycle and reuse materials remaining from a project for another project.

(4) **Safety of workers and workplace:** Considering the hazardous nature of the construction industry, the safety of workers is one of the significant concerns of project contractors. Over time, stakeholders in the construction industry have realised that minimizing injury on the construction site can be attainable through off-site construction operations. Off-site construction allows workers to work in a spacious environment with minimal collision. This is made possible through prefabrication techniques where a controlled environment where risks on construction sites such as sliding and electrocution scaffolding are minimized. Hence, construction workers are free from injuries that might occur due to an accident of large vehicles moving on the construction site. Also, there is a better set-up plan, and the potential of workers getting injured is minimized due to fewer workers on site.

(5) **The solution to the shortage of local skilled labour**: Shortage of skilled labour has always been a significant challenge in the construction industry. Even though several initiatives and research have been conducted to resolve this issue, the challenge still prevails in the industry. However, the invention of the modular construction system offers advanced technology for construction, where less labour is required.

(6) **Weather and Climate:** Prefabrication construction occurs in an enclosed off-site factory environment that is hardly affected or delayed by weather and environmental conditions such as rain, intense sun, cold, snow, storms etc. (Navaratnam et al., 2019). On the other hand, the on-site construction process is susceptible to diverse environmental conditions, which can cause delays and eventually the project's outcome.

(7) **Product Quality:** Prefabrication entails manufacturing significant off-site construction components, which allows for the quality outcome of building construction. The use of industrialized components for prefabrication gives room for a detailed process that ensures conformity with standards towards

achieving better quality. Better quality in prefabrication is achieved as the manufactured components help minimize risks, waste, and other dangers on construction sites that might affect the quality of the project output.

(8) **Aesthetic Appeal:** Prefabrication makes the building of any shape possible with different façades and beautification.

5.5 Risk of Adopting Prefabrication in Construction

Despite the benefits of prefabrication, it comes with some risks and constraints. It is pertinent to address the risks for better implementation in the construction industry. According to Sabrina and Tamer 2019; Lin et al., 2019; Wuni et al., 2019, the risk of prefabrication can be classified into the following:

(a) **Structural risks:** In recent times, discoveries have shown that natural change risks resulting from climatic conditions such as earthquakes, flooding, landslides, typhoons, cyclones are divulging some structural risks in the construction field (Lin et al., 2019). Modular construction comprises completed integrated modules such as fittings, fixtures, and finishes that may be susceptible to environments that encounter natural disasters. As a result, these natural events present a significant risk to prefabricated structures, resulting in the project's structural failure.

(b) **Ergonomic risks:** Ergonomic risks are physical conditions that may pose injury or risk resulting in musculoskeletal disorders. The factors that constitute ergonomic risk can cause biomechanical strain on the employee. Construction workers are exposed to ergonomic risk because there is still manual handling of modular components such as screwing, unloading, welding and inspection in prefabrication. Manual activities in prefabrication which involves heavy masses of modular components, cause risks to workers' safety. Manual handling and operations during modular construction and on-site assembly are guidelines to ensure protection from ergonomic risks. There have been reported cases of construction workers in the United Kingdom with severe fatigue. They manually inspected, unpacked, lined up, unfastened, screwed, and welded modules and enabled crane lift upon the arrival of modules to a construction. Again, the manual handling, inspection, unloading, screwing and welding of the heavy modular elements upon arrival to the construction site exposes workers to injuries and risk of work-related musculoskeletal disorders (WMSDs).

(c) **Inflexibility in Changes:** Prefabrication components are already fabricated based on agreed design and specifications of the clients. Consequently, any form of error or adjustment intended to be made is often difficult to effect on the prefabricated construction. Also, considering the lifespan of a building (which may extend to 80 years), the modification of the indoor building may be required as a result of development, changes in taste and lifestyle. However, this may not be possible in prefabrication buildings.

(d) **Higher Initial Capital:** Although prefabrication has proven to be a cost-effective construction system, the implementation demands a high initial capital. The application of prefabrication involves the initial purchase of plant and machines, land for manufacturing plant, modular storage warehouse, software and cranes. The high initial capital cost constitutes disadvantageous recipes for large risk spectra (Wuni et al., 2019). The severity of this risk is felt more by economies with diseconomies of scale, where the producers might find it challenging to break even to recoup the initial capital invested.

(e) **Limited Customization:** Modular structures are indeed customizable according to the needs and expectations of the clients. But it is also true that there is a limit to how much customization can go into it. Although the prefabrication component kits can be custom-made for different projects, there's a limitation to the design of each component. Only limited types of components are available for various projects. Hence, if you are looking for a highly personalized and customized project, modular construction may not be the correct type of construction for that purpose.

(f) **Damage in Transit:** Prefabrication involves the transportation of construction modules and equipment from the place where it is prefabricated to the construction sites. Some of these prefabricated modules are large and delicate, posing a significant risk of transit damage. Hence, additional materials are required to be joined with the modules to minimize damage during transportation (Sabrina and Tamer, 2019).

5.6 Relevant Skills for Prefabrication Application

Compared to other industry digitalization and technological advancements, the construction industry has not been fully developed because of its cost and labour intensive setbacks (Dallasega et al., 2018; Woodhead et al., 2018). The skills shortage has always been one significant issue faced by the construction industry. For the application of prefabrication, professionals such as architects, engineers, quantity surveyors and few manual workers such as bricklayers, carpenters are required to possess specific skills. More importantly, prefabrication requires a high level of skills in planning, design, manufacturing, installation of prefabricated modules to achieve a quality prefabricated construction. Consequently, the following skills are relevant for the implementation of prefabrication. According to Buddhini and et al., (2019), skills needed to carry out prefabrication successfully include the following:

(1) **Designing Skills:** Designing skills is an essential skill in construction and prefabrication/modular construction. It involves reading and producing a set of drawings from which the building will be built. This also involves: design integration, understanding buildability, knowledge and expertise, business management skills (such as generating new business models), and achieving repeat orders of different technologies to deliver a prefabricated building. Majorly, prefabrication is an offsite construction demand skill in Design for

Manufacture and Assembly (DfMA). (Ginigaddara et al. 2019). This skill is significant in the success of the modular project.

(2) **Digital Skills:** Prefabrication construction, categorized as offsite manufacturing of building components, uses automated production techniques requiring digital skills. These encompass information technology and programming. These skills are essential to use digital devices, access and manage information and communicate effectively. For example, digital skill is essential in designing an off-site building.

(3) **Data Analytics Skill:** These are analytical skills in performance parameters, including structural, energy efficiency, material strength, and cost.

(4) **Planning Skill:** Planning is essential for any project to be carried out successfully, but in prefabrication, planning is not just a prerequisite to start with, but what the whole project is about. It takes professional ability to think about activities, materials and labour needed to carry out a prefabrication project. The ability to plan well could be the difference between profit and loss. Well-grounded planning skills in prefabrication would entail creating a blueprint of a project from start to finish, and prefabrication can only take place after the project has been completed on paper. Prefabrication rides on detailed planning and organizing carried out before work starts. Planning skill directs all other skills needed for prefabrication. Planning skills would be reflected in the material handling of any construction project from start to finish.

(5) **Installation Skill:** The installation stage is one of the most crucial aspects of any construction or manufacturing setting. Successful installation tends to be a successful project. Prefabricated components would not arrange themselves: it needs professionals to get it fixed onsite. A prefabricated building module includes the components of a conventional multi-story house project and the well-designed installation procedure. Each module has been configured to provide a range of room and dual-room configurations, and without proper installation, tech know-how, erection would never be possible. A good installation skill all-round would encompass the ability of a person or firm to be compliant with safety and building integrity, environmentally sensitive, and must pay immense attention to details.

(6) **Automation Skill:** Automation has to do with the ability of a process to be carried out with little or less human intervention. Automation is similar to the manufacturing industry, where parts are coupled and arranged by an installed automated process with little or no human intervention or supervision. Automation skills in the construction industry or instead in prefabrication have to do with creating a process of arrangement of building installation with little stress and human intervention. Automation saves time and energy, and prefabrication is a product of an automation process. A well-grounded automation skill would involve the ability to understand development methodology, to communicate skills through the building process effectively, have an idea of computer programming and excellent acute attention to detail.

(7) **Collaboration:** Prefabrication cannot be done just by having one person or a set of people with just one skill. In prefabrication, different professional skills

will be needed across every unit, and all these experts need to collaborate as a team for success to be recorded. It's not a one-person job; it is a job that requires different people having different skills, playing their parts well and synergizing with one another for the project optimization.

There is a critical need for the construction personnel to possess digitalisation, automation, and industrialisation skills to cope with the smart city. The current off-site construction with the presence of the internet of things, cloud computing, data analytics, cybersecurity etc., demands the inclusion of higher-level skills. Hence, creating the need for up-skilling and reskilling to match up to modern technological requirements (Buddhini et al., 2019; Woodhead et al., 2018).

5.7 Summary

Prefabrication/Modular construction, which has existed for a long time, is currently gaining much recognition as a building technique in the construction 4.0 era. Stakeholders in the construction industries such as professionals, contractors, clients and even the government are beginning to adopt this building technique as a solution to some of the challenges, such as the shortage of building faced in the industry. The introduction of new technology in the construction industry has improved prefabrication, thereby creating comprehensible building structures within a little space of time. This building technique applies to almost all construction categories, including residential, commercial, bridges, stadiums, and grandstands. Adopting this building method offers several benefits, which makes it a better option than conventional construction. These benefits include -saving, cost and waste reduction, promoting the safety of workers and workplace, and solution to the shortage of skilled labour and better-quality construction project. Similarly, some risks such as higher initial capital, inflexibility in changes, damage during transit, structural and ergonomic risk, and the implementation of this building method. However, identification and management of the risks make the prefabrication method more effective. Furthermore, the implementation of prefabrication and Modular in the current digitalized construction (Construction 4.0) demands a high level of skill comprising soft skill and technical skills. The soft skill includes planning, communication, interpersonal, collaboration, negotiation, team skill. On the other hand, the technical skills include designing, digital, automation, programming and installation. Therefore, the need for up-skilling and reskilling of the construction workforce becomes necessary and inevitable.

In summary, prefabrication will make houses and buildings less cumbersome to erect as prefabrication will make building projects faster, thereby tackling the shortage of housing caused by an increase in population around the world. Buildings would not just be allowed to spring up without proper documentation and certification from professional government agencies. Policies and regulations will be channelled towards safe and sustainable prefabricated housing projects as it has done with conventional building requirements.

References

Alireza, B., Meltem, V. K., & Omid, R. B. (2015). Using prefabrication systems in building construction. *International Journal of Applied Engineering Research, 10*(24), 44258–44258. https://www.researchgate.net/publication/296639797.

Ankit, S. (2016). Role of prefabrication in the construction industry to improve sustainability and economy of civil engineering projects. 1–23. University of Illinois, Urbana-Champaign: Research Gate. doi:https://doi.org/10.13140/RG.2.1.3798.8246

Anuradha, D. J., & Dhawade, S. M. (2015). Benefits of prefabricated building material over on site construction-a review. *International Journal of Research in Engineering, Science and Technology, 1*(8), 195–202.

Buddhini, G., Srinath, P., Yingbin, F., & Payam, R. (2019). Skills required for offsite construction. *CIB World Building Congress* (pp. 1–12). Hong Kong SAR, China: Research Gate. https://www.researchgate.net/publication/333990891

Chinmay, R. K. (2019). Methodology of 3S construction method "prefabrication technology." *International Research Journal of Engineering and Technology, 6*(12), 1–5.

Dallasega, P., Rauch, E., & Linder, C. (2018). Industry 4.0 as an enabler of proximity for construction supply chains: A systematic literature review. *Computers in Industry, 99*, 205–225.

Fred, E. B., Jin-Hee, K., & Jun-Tae, K. (2016). Performance of modular prefabricated architecture: Case study-based Review and future pathways. *Sustainability, 8*(558), 1–16. https://doi.org/10.3390/su8060558

Ginigaddara, B., Perera, S., Feng, Y., & Rahnamayiezekavat, P. (2019). Skills required for offsite construction, conference proceedings, CIB World Building Congress 2019, 17-21st June, Hong Kong, China

Hearn, J. (2018, June 13). *A shot history of prefabrication | History, Desgin | Prefab Museum.* Retrieved from: https://www.prefabmuseum.uk/content/history/short-history-prefabrication. A short history of prefabrication.

Hopkins, A. S. (2014). Assessment of modular construction utilization across all campuses of the University of North Carolina system. *M.S. thesis, University of North Carolina.* North Carolina, USA.

John, C. (2020). *On the West Coast, prefab gains ground for speedier construction.* https://www.bdcnetwork.com/west-coast-prefab-gains-ground-speedier-construction

Lauren, H. (2015). *High-Speed High Rise: The Earthquake-Proof Skyscrapers Built in 15 Days.* http://www.wired.co.uk/article/high-speed-high-rise

Lin, K., Lu, X., Li, Y., & Guan, H. (2019). Experimental study of a novel multihazard resistant prefabricated concrete frame structure. *Soil Dynamics and Earthquake Engineering, 119*, 390–407.

Lida, T., Liming, L., Kang, L., Jiao, G., & Mind, L. (2018). Study of detailing designing of precast concrete frame structure. *IOP Conference Series: Earth and Environment Science, 128.* https://doi.org/10.1088/1755-1315/128/1/012085.

Martin, A. F. (2019). Modular construction in multi-storey buildings. *Master's degree in Industrial Engineering specialized in structures and construction.* Padua, Italy: University of Padua. http://hdl.handle.net/10810/36709

Muhammad, A., Sohail, M., & Sumra, Y. (2017). Performance evaluation of cost-saving towards sustainability In traditional construction using prefabrication technique. *International Journal of Research in Engineering and Science*, *5*(5), 73–79.

Mohiuddin, A. (2015). *Accelerated bridge construction*. Butterworth-Heinemann.

Navaratnam, S., Ngo, T., Gunawardena, T., & Henderson, D. (2019). Performance review of prefabricated building systems and future research in Australia. *Buildings*, *9*(38), 1–13. https://doi.org/10.3390/buildings9020038.

Polser. (n.d.). *Fiberglass and its Use in Construction*. https://polser.com/en/frp/fiberglass-and-its-use-in-construction#:~:text=Fiberglass%20glasses%20are%20formed%20into,in%20construction%20and%20construction%20sector.

Radan, T. (2017). Advantages of precast concrete in highway infrastructure construction. In *Creative Construction Conference. 196*, (pp. 176–180.) Primosten, Croatia: ScienceDirect: Procedia Engineering. doi:https://doi.org/10.1016/j.proeng.2017.07.188.

Redshift, V. (2019). *The History of Prefabrication, From Roman Forts to Modern Modular Housing*. https://www.autodesk.com/redshift/history-of-prefabrication/.

Sabrina, S., & Tamer, B. (2019). Analyzing methods of prefabrication and their application in the construction of habitat for humanity housing. *Journal of Architecture and Civil Engineering*, *4*(1), 42–49.

Tharaka, G., Roshan, K., Priyan, M., & Tuan, N. (2016). Prefabricated construction technologies for the future of Sri Lanka's construction industry. In *The 7th International Conference on Sustainable Built Environment* (pp. 1–8). Kandy, Sri Lanka: Researchgate. https://www.researchgate.net/publication/312032697.

Tharaka, G., Tuan, N., Priyan, M., & Shan, K. (2017). Performance of multi-storey prefabricated modular buildings with infill concrete walls subjected to earthquake loads. *Concrete in Australia*, *43*(3), 51–58. https://www.researchgate.net/publication/319933150

The State of Victoria Department of Environment, Land, Water and Planning, (2015). 383 La Trobe Street, Melbourne Officer Assessment Report. https://www.planning.vic.gov.au/__data/assets/pdf_file/0013/104035/21-PA1600102-383-La-Trobe-Street,-Melbourne.pdf.

Woodhead, R., Stephenson, P., & Morrey, D. (2018). Digital construction: From point solutions to IoT ecosystem. *Automation in Construction*, *93*, 35–46.

Wuni, I. Y., Shen, G. Q., & Hwang, B. (2019). Risks of modular integrated construction: A review and future risk direction. *Front. Eng. Manag.*, 1–18. https://doi.org/10.1007/s42524-019-0059-7

Zhong-Lei, W., Hou-Cai, S., & Jian, Z. (2019). Risks in prefabricated buildings in China: Importance-performance analysis approach. *Sustainability*, *11*, 3450. https://doi.org/10.3390/su11123450.

Chapter 6
3D Printing/Addictive Manufacturing

6.1 Introduction

3D printing, also referred to as Addictive Manufacturing (AM) is categorized as one of the Construction 4.0 technologies. Technology is considered one of the fastest-growing technology in almost all industries, including construction. This technology involves the construction of a Three-Dimensional solid object from a digital model. This technology includes the manufacturing of building modules with a computer-aided diagram (CAD). It is primarily useful for prototyping and the manufacture of geometrically composite components. 3D Printing started in the early twentieth century with the first-ever 3D printer invented in 1984, evolving from automated production. It had its first application in the manufacturing and automotive industries. Recently, 3D has become one of the fastest-growing technologies, with its applications covering several industries. The application of this technology has proved effective and beneficial in industries such as medical, aerospace, and construction, thereby providing the opportunity for mass customization and complex shapes that cannot be produced in other ways.

Historically, 3D Printing, termed Rapid Prototyping (RP) technology, was conceived in Japan in the 80's. In 1984, a method of 3D printing known as stereolithography (SLA) was invented by an American Inventor. Stereolithography (SLA) is a liquid photopolymer resin contained in a reservoir, as shown in Fig. 6.1.

Subsequently, the 3D model created from digital files is printed into a solid and physical shape with one layer, UV lasers which are used to activate the resin to solidify. In architectural firms, the printing tabletop scale model represents one of the earliest uses of 3D printing. These models enhanced the design process and were a valuable tool for planning the marketing of building projects. Today, stereolithography (SLA), one of the early invented 3D models, is still one of the most popular 3D printing technologies. Other common additive technologies include selective laser sintering (SLS), fused deposition modelling (FDM) and direct metal deposition (DMD) (Ellis, 2018). Over the years, the capabilities of 3D printing have extended beyond its initial potentials. For example, a system of 3D printing called

O. Adepoju et al., *Re-skilling Human Resources for Construction 4.0*, Springer Tracts in Civil Engineering, https://doi.org/10.1007/978-3-030-85973-2_6

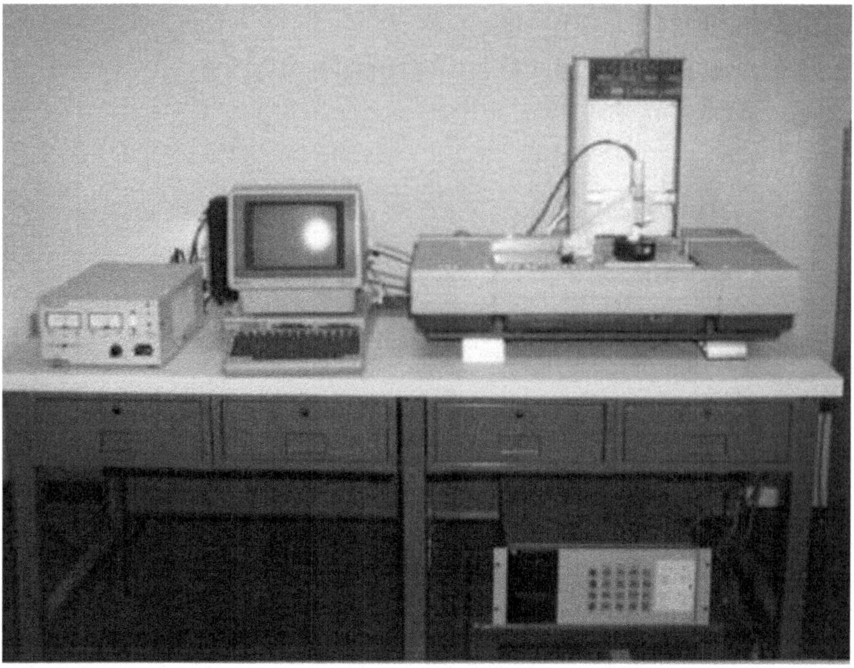

Fig. 6.1 First Invented stereolithography (Giacomelli & Smedberg, 2014)

Contour Crafting is designed to print buildings in place. Recently, the automated and accelerated process of 3D printing is also auspicious for civil structures such as buildings and bridges, which requires extensive labour.

In recent times, the application of 3D printing has brought about a significant change in the construction industry. The adoption of 3D, known as Construction 3D Printers, has enhanced the building process, such as a higher speed of construction and assembling, less use of labour, less waste of resources and materials. The 3D machines, alternatively referred to as construction 3D printers, are technologies used to build houses by depositing paste-type material, either concrete or earth materials, pushed through a nozzle layer by layer to print buildings in 3D (Cherdo, 2020). The 3D digital model that sets the foundation of any 3D printing process is created using various 3D software programs. This model is cut into layers according to design and further converted into a readable file by the 3D printer. Globally, several constructions are built with 3-dimensional printing technology. The 3D Printing Process in Construction involves inputting appropriate construction data such as plans, maps, dimensions, and designs into an automated machine that generates a specific physical output. The three-dimension printer uses a stereolithography procedure which involves the laying of a product layer by layer pattern. Building components such as walls, partitions and beams can be produced with an automated 3D printer on

Fig. 6.2 3D printing process (Mehmet & Yusuf, 2017)

the construction sites, while other building components can be installed using the traditional methods. In construction, 3D printing involves the following five stages:

- 3D Model Computer Designing which consists of the preparation of graphical design with the help of CAD software;
- Conversion and exporting of model into Stereolithography/standard tessellation language (STL) Format which is adapted to work;
- Processing to decompose into slices and required adjustments are made to the designs and orientation;
- Reading and laying down designs into successive layers of 3D object printing structure, which is generated by filling up of the cartridges with the necessary materials such as powders, liquid, binders and sheet;
- Production of the structure, which is the final object produced.

These stages listed above are graphically depicted in Fig. 6.2 below make it possible for any shape to be produced using 3D printing.

Furthermore, in construction, the 3D materials are vital as they determine the structure's quality. Several materials can be utilized for 3D printing. These include: concrete, silver, steel, glass, titanium, photopolymers, stereolithography, polycarbonate, Acrylonitrile Butadiene Styrene (ABS) plastic, Polylactic acid (PLA). These materials can be mixed with other substances to get a perfect solution, such as concrete mix with recycled material, fibre, sand and geopolymer. 3D printing has provided a cheaper, faster and more sustainable alternative to construction in the world today.

6.1.1 3D Printing Materials

Several materials can be used for 3D printing. However, these are selected based on the method of 3D printing to be adopted and requires the identification of 3D techniques. The common materials can be classified into Cement, Polymers, Metals and Ceramics.

(i) **Polymers:** Polymers for 3D printing/additive manufacturing comes in the form of filaments, thermoplastic, reactive monomers, resin or powder. Polymers are the most-used materials in 3D printing. This results from its availability, low cost, diversity and ease of adoption in printing processes (Alberta & Paolo, 2020; Ngo et al., 2018). More importantly, it is widely used because of its cost-effectiveness compared to traditional methods. Polymers utilize

3D techniques, including laminated object manufacturing, fused deposition modelling, stereolithography, selective laser sintering, 3D bioprinting, and Inkjet printing. Each of these has techniques designed to process only a specific polymer. In the industrial field, the most-used polymer technique is the Photopolymer resins used for stereolithography due to the perceived possibility of excellent achievable accuracy during the building process (Ngo et al., 2018). The only concern in the use of polymers is mechanical properties and pure polymer products built by 3D printers, which are used explicitly for conceptual prototypes due to the innate lack of strength and functionality (Alberta & Paolo, 2020; Ngo et al., 2018; Parandoush & Lin, 2017).

(ii) Ceramics: The main methods of 3D printing ceramics are inkjet (suspension), powder bed fusion, paste extrusion, stereolithography and selective laser sintering (SLS) of powder. Inkjet is deemed the primary method of making dense ceramic samples that may not need post-treatment (Travitzky et al., 2014; Ngo et al., 2018). Using Ceramics in 3D printing has the advantage of controlling the porosity of lattices. This advantage ushered in the benefit of developing advanced lightweight materials tailored for different applications (Withell et al., 2012; Ngo et al., 2018). The ceramic AM processes have rapidly evolved in recent years; however, in many cases, the mechanical properties of manufactured parts do not reach the desired values (Alberta & Paolo, 2020; Wang, Dommati & Hsieh, 2019).

(iii) **Metals:** Metals and alloy materials are increasingly growing in use for 3D printing technology. Metals in 3D printing are majorly processed by melting metallic feedstock made from powder or wire using an energy source such as a laser or an electron beam. Melted are commonly used materials that are transformed layer by layer to form a solid part through several techniques. The methods for transforming metals to printing metals are powder bed fusion (PBF) and direct energy deposition (DED). Powder bed fusion technologies possessed good mechanical properties and can produce components with complex shapes of high accuracy. However, one of the shortcomings of these technologies is that it is slow and as a result mainly used for small parts (Ngo et al., 2018). Nevertheless, in recent times, other techniques such as cold spraying, friction stir welding, binder jetting, direct metal writing and diode-based processes have been developed (Ngo et al., 2018). These processes can achieve higher accuracy or speed. Examples of metallic materials used in 3D printing include: steels (austenitic stainless steels, maraging steels, precipitation hardenable stainless steels, and tool steels), titanium and its alloys, aluminium alloys, and nickel-based alloys (Herzog, Seyda, Wycisk & Emmelmann, 2016; Ngo et al., 2018; Alberta & Paolo, 2020).

(iv) **Titanium:** Titanium alloys are the most-used alloys in additive manufacturing. The first type of alloy processed and used in several industries, such as automotive and aerospace, is 3D Steel. Among the processable steels of 3D printing, the most explored are the stainless, tool, maraging and precipitation hardening (PH) steels. We also have aluminium alloys, which has shown enormous success in 3D printing/addictive manufacturing. However, this type of alloy is limited in use due to the tendency of cracking due to solidification during 3d processing. This use of this material presents several benefits, such as the ability to control interspatial content and produce customized and complex 3D printing designs.

(v) **Concrete:** Generally, concrete is the most used material in the construction process around the world. Concrete material has been integrated into 3D Printing. There are several compositions and methods available for concrete blend and technology that can be adopted. The composition and the concrete mixture is very important as it determines the strength and end product of the 3D printing. Fibre-reinforced concrete could also be printed using 3D printing technology. 3D printed fibre-reinforced concrete composites benefit from increased flexural strength and controlling fibre orientation in a printed structure compared to traditional fibre-reinforced concrete (Hambach & Volkmer, 2017; Ngo et al., 2018).

6.1.2 Methods of Three-Dimensional (3D) Printing Technologies in Construction

There are several types of 3D printing technologies that can be used to process different materials to create a final product. There are four main methods of 3D printing technology in construction. These include the following:

(i) **Contour Crafting (CC):** Contour Crafting is a layered additive fabrication technology developed by Behrokh Khoshnevis of the University of Southern California (Khoshnevis, 2004). CC is computer controlled to exploit the superior surface-forming capability of troweling to create smooth and accurate planar and free-form surfaces. Technology is considered the most promising 3D printing technology in the building industry as it gives a great opportunity to automate the construction process. This technology makes it possible to print a whole house directly on-site with the use of 3D printers. The Contour Crafting of 3D printers shown in Fig. 6.2 makes it possible to construct complex structures on sites with crane and robotic arm aid. This process is conducted on-site as the materials are poured progressively into layers. The idea behind the invention of the CC technology is to create a printer that will have one or few nozzles that are moving on two parallel lanes installed at the construction site, separated a few meters from each other and broader than the width of the building. The material is thrust through the nozzle and laid down in the shape of empty blocks, with a crosswise pattern inside to ensure desired stiffness and strength (Izabela et al., 2016). Several other methods of CC machines have been designed to construct both small and large-scale structures effectively. Figure 6.2 display A 3D printed castle of Andy Rudenko's garden produced by the Contour Crafting technology. The design of these models involved a mix of cement and sand printed on a single run. However, the towers were printed separately and assembled in the building (Figs. 6.3

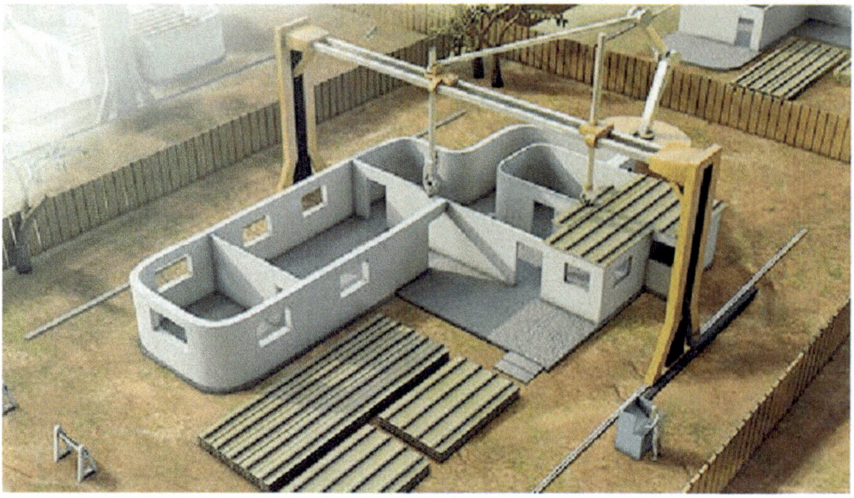

Fig. 6.3 Contour crafting 3D printer (Sagheer et al., 2018)

and 6.4).

(ii) **D-Shape:** D shape is a 3D invention developed by an Italian engineer Enrico Dini in 2004. This technology is the only original large-scale 3D printing initiative used to digitally design and construct a multinational building structure. This technique made it possible to construct a structure at any shape from the basement up to the roof at once. The D shape technology is essentially a large 3D printer that uses binder jetting technology. The opportunity of a large gantry structure of the D Shape to move the printer creates an alternative type of concrete. This is achieved by selectively applying a liquid binder on top of layers of powder material consisting of a cement-sand blend. Following the application of the binder, the powder material solidifies, while the residual unbound material remains as a support for the hardened parts.

Fig. 6.4 **a** First structure printed in-situ; **b** printing progress (Izabela et al., 2016)

One of the significant features of the 3D printer is that it allows the printing of 3D models directly from a file into concrete-like shapes with a high degree of freedom. There are several opportunities of the D shape 3D printing that avails avail contractors faster and cheaper methods of construction; developers offer of customized building, well suited to produce unique pieces and building components and best suited for off-site prefabrication of various highly complex construction elements. Figure 6.5 shows D-shape printing technology.

(iii) **Concrete Printing:** A group of people developed Concrete Printing technology at the University of Loughborough, United Kingdom. Concrete which is the most used construction material in the construction industry faces several challenges. The invention of 3D concrete printing has made construction work easier and resolve many challenges faced with the conventional use of concrete. The use of 3D concrete printing avails opportunities such as creativity and geometrical freedom for architects, improve workers 'safety, facilitate construction, thereby reducing delay. However, the concrete printing technology has been developed to retain 3D freedom and has a smaller deposition resolution, which allows for greater control of internal and external geometries. The 3D concrete printing can be categorized into two. These include the factory and the site. The factory creates a centralized location with a controlled environment, making it possible to produce construction elements without interruption. The project team also can inspect the production anytime

Fig. 6.5 D-shape printing technology (Hamad et al., 2018)

so desire. On the other hand, the 3D concrete printing -site takes place at the construction site, thereby availing the opportunity to produce larger units and less transportation. Nevertheless, this method demands security of intellectual property, mobilization, land and climate limitation.

(iv) **3D Printing Metal:** In construction, metal addictive manufacturing has been developed for structures that can withstand more pressure. This is a laser-based technology that uses powdered metal, which fixes together particles on a powder bed with layers of metallic powder distributed by the technology. This technology combined a manufactured robot with a welding machine to turn it into a 3D printer. In construction, metallic materials commonly used include aluminium, carbon steel and stainless steel. The metal 3D printing enabled through lightning nodes makes it possible to achieve an efficient, lightweight structure that might be costly and time-consuming using traditional methods. Typical examples of metal 3D printing in construction include the façade node and lighting node, as shown in Fig. 6.6.

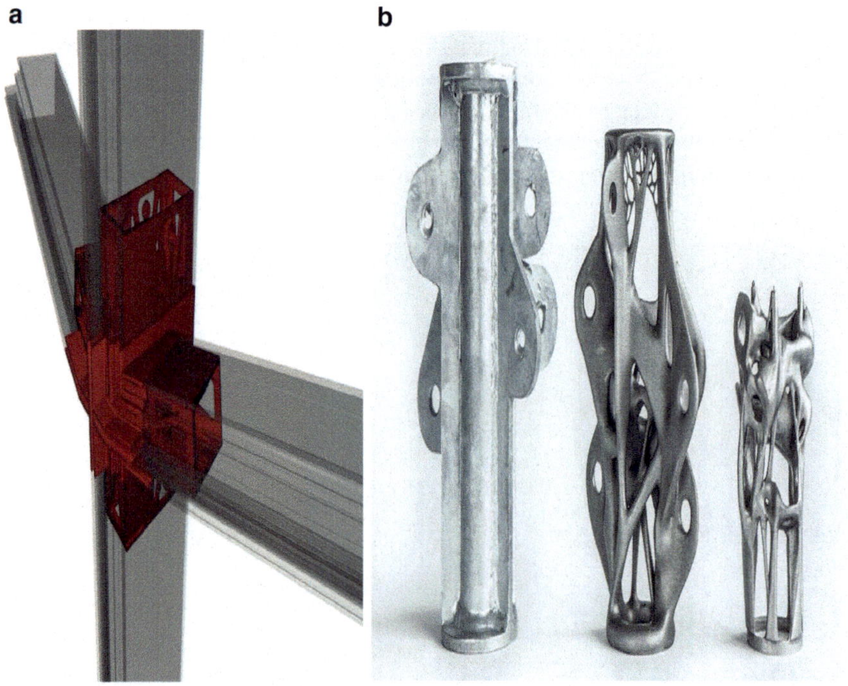

Fig. 6.6 **a** 3D Metal lighting node **b** 3D Metal façade node (Buchanan & Gardner, 2019)

6.2 Application of 3D Printing to Construction

Since 2004, when the first 3D wall was built, 3D technology has experienced dynamic advancement and applied in numerous forms in the construction industry. These applications are explained below.

6.2.1 3D Architectural Design

Building design lays the foundation of a quality building project and is considered a key element by architectural and other building professionals. Clear and accurate building designs help clients to visualize their ideas vividly. This is made possible through the 3D printer. 3D printing has transformed architectural designs, thereby making it possible to design a building model accurately within few hours. It avails architects and designers the opportunity for added and more innovative designs, early detection of mistakes before construction, saving cost and reducing human error. Also, complex shapes can be printed, and detailed designs can be made. This may include interior, complex facades, landscape, environmental elements, towns, and cities that can be designed. These opportunities have replaced the traditional method of designing, which is tedious, expensive, and depends intensely on the skilled artisan. Fig depict an architectural design for a structure and light network (Fig. 6.7).

Fig. 6.7 A 3D printing architectural design structure (left) and light network (right) (Raspalli & Banon, 2017)

Fig. 6.8 The three-dimensional modern printer (Shakir, 2019)

6.2.2 3D Printed Houses

3D or Addictive Manufacturing has become an indispensable technology for the construction industry. This technology which consists of a head robot that moves in and out in the form of a paste-like substance known as the jetting of the printer ink shown in Fig. 6.8, has made it possible to construct notable buildings structures with great designs within hours. In the same pattern of creating 3D objects, 3D houses have become a reality. These houses are printed using reinforced and enriched concrete mixtures such as sand, cement, fibers, geopolymers. Concrete is mixed with additives to ensure the hardening of the 3D printed house. Also, several building components which form a building structure can be printed using 3D printing technology. The 3D building components include walls, stairs, and roofs. A typical example of a 3D printed house is the Apis Cor printed house in Russian (Fig. 6.9) which was constructed 24 h.

6.2.3 3D Office

3D printing technology has also been applied to office construction using concrete, glass fibre and plastic. The innovation of the 3D printing construction office is a new design that offers a solution to one of the challenges of inadequate building. 3D printing allows designing the interiors of the office structures, such as furniture and partitions. A typical example of a 3D office is the first world 3 Dimension Printing Office in Dubai, the United Arab Emirates, as shown in Fig. 6.10. This office was constructed to fulfil the vision of the leaders in Dubai and serve as a place where people worldwide can converge. The 3D printing is fully functional with all required facilities.

Fig. 6.9 Apis Cor 3D printed house in Russian (Mehmet & Yusuf, 2017)

Fig. 6.10 **a** Side view **b** front view, 3D printing office in Dubai (Mehmet & Yusuf, 2017)

6.2.4 3D Bridge and Canal

3D printing can be applied to the construction of bridges and canals. Recently, a fast-growing company known as Dutch MX3D specialising in 3D printing revealed a metal bridge crossing an ancient canal in Amsterdam, the Netherlands' capital. The bridge's design was carried out at the company's workshop using a 3D printing robot. The MX3D bridge confirms that metal 3D printing is probable at a measure appropriate for application in construction. The mental 3D printing shows a possibility of constructing an efficient, lightweight structure such as the bridge that could

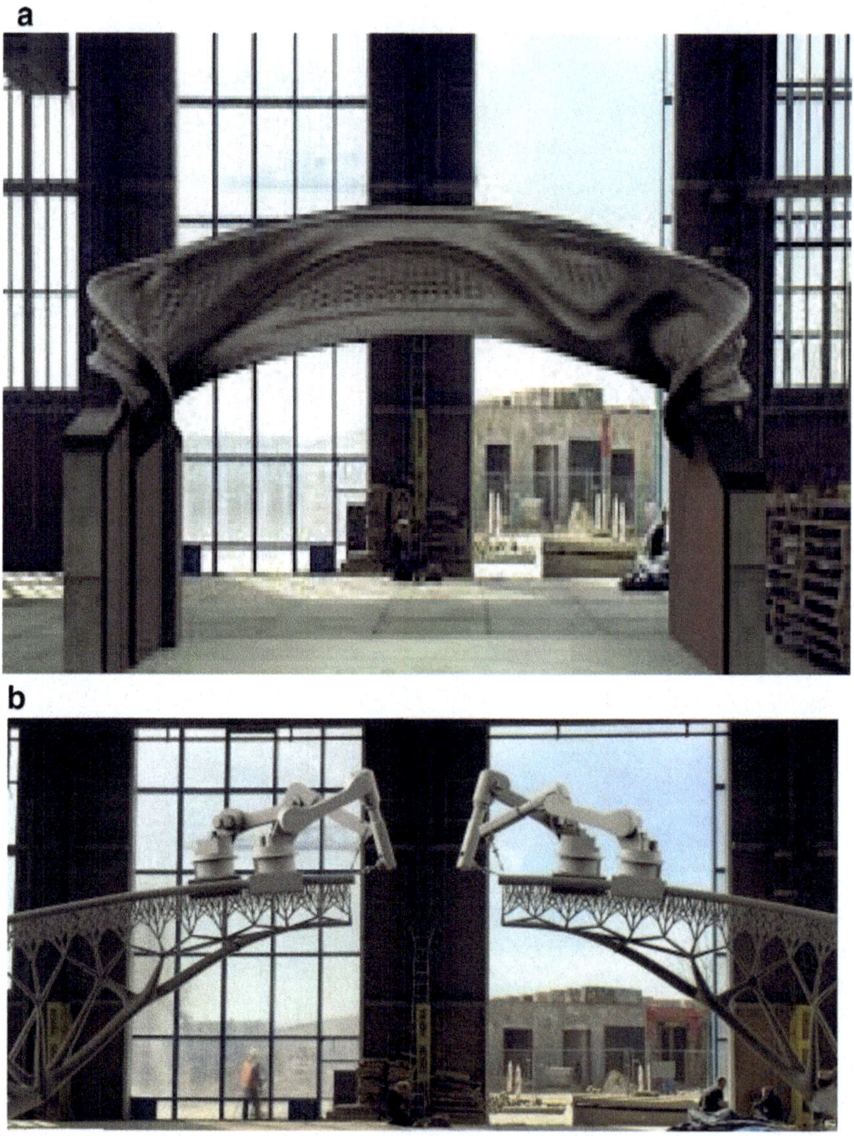

Fig. 6.11 **a** MX3D bridge in Amsterdam (Hossain et al., 2020) **b** MX3D Bridge in Amsterdam (Hossain et al., 2020)

take a long time and be costly to manufacture using the traditional farming method. Figure 6.11 shows the completed MX3 bridge in Amsterdam, Netherlands.

Fig. 6.12 3D-printed stormwater collectors (GaudilliEre et al., 2018)

6.2.5 *Stormwater Collectors*

3D can also be applied in the construction of stormwater collectors. The stormwater collector is a system designed to capture, retain and stormwater for valuable use. The stormwater is stored in ponds, land surfaces and soak into the soil. Stormwater can be used for domestic use such as drinking, washing, cooking and public use such as farm irrigation, firefighting etc. In the case of Lille and Roubaix construction project in France, a 3D stormwater collector was produced and installed on-site by Xtree, shown in Fig. 6.12. This was conceived in a workshop to explore the potential of three-dimension concrete printing (3DCP) in public works. The adoption of 3D printing for producing the stormwater took two days instead of two weeks. As a result, this offered the benefit of speed, convenience and a radical reduction in onsite involvement. Also, 3D offered the opportunity for several designs, workflow integration.

6.3 Case Studies of 3D Printing Applications in Construction Projects and Resultant Benefits

6.3.1 Case Study One: Apis Cor Printed House in Russian

Apis Cor printed house situated in Mosco was designed in 2016 by a Russian company specialising in developing mobile construction 3D printers. The 3D printing technology makes it possible to design a whole building on site. The Apis Cor printed house, which comprised 400-square foot concrete walls and foundation and facilities such as windows, fixtures, and furniture, was constructed 24 h from scratch. The total cost implication of the entire project, including materials, foundation, walls, roofing, windows, interior and exterior, amounted to $10,134. This printed house was built entirely on the site with the use of the 3D printer alone. The final construction of the house can be described as habitable and comfortable with equipment such as a sofa, fridge and Television. Naturally, building a house takes months. Thus far, the Apis Cor 3D printed house was completed at a lower cost within a day that ordinarily would have taken an average of two months to construct using the traditional construction method. The resultant benefits of using 3D printing include speed, low cost, less personnel and human error due to automation.

6.3.2 Case Study Two: 3D Printed Office in Dubai

The world's first 3D printed office in Dubai was designed for the United Arab Emirates to serve as a conference space for parties from all over the world. It also serves as a management office for museums of future projects that will help deliver SMART technologies to the Emirates. This structure, a 2,000 square feet office and exhibition space was printed with a 20 feet tall 3D printer. The components of the 3D printed structure were produced in China and shipped to Dubai for installation. This office structure was constructed using 3D printed concrete, plastic and glass fibre in nineteen (19) days to print and install with a total cost of $140,000. Due to the high level of automation, the structure was completed with the assistance of eight personnel; one personnel oversee the printing process, and eight were involved in installing the building components on-site. The 3D printed office has various facilities such as water, telecommunications and air conditioning systems. The resultant benefits of adopting 3D printing in constructing this project include speedy delivery of the project (19 days), reduction in the cost of construction, labour (50% to 80% reduction in labour cost), efficient utilization of materials (30% to 60% reduction in waste).

6.3.3 *Case Study Two (3): Freeform Truss Shaped Pillar in Aix-En-Provence, France*

The Freeform pillar is a 4 m-high column situated at the sports centre of a school in Aix-en Provence, France. This pillar was made of integrated concrete 3D printing that covers a part of the school's playground. The XtreeE 3D printing technology was used for fabricating the architectonic elements, which made it possible to construct unusual geometrics with curved and fine elements. The pillar was printed at XtreeE in Paris, France, the casting and integrated connectors were produced in Germany and the final assembling of the structure on-site in Aix-en Provence. 62.5% is gained from the total cost of production using 3D printing compared to traditional manufacturing. In summary, the resultant benefits of using 3D/Addictive Manufacturing include: materials saving, reduction in labour and safety (through the elimination of on-site manual construction process), time-saving (through reduction of some steps in the traditional construction process), and workflow between project stakeholders. Also, the adoption of 3D printing creates the possibility to achieve the complexity of the composition of the 3D shape of the pillar. (GaudilliEre et al., 2018).

6.3.4 *Case Study Four (4): Ten (10) 200 Square Meters Houses in China*

A Chinese company named WinSun Decoration Design Engineering Company printed ten full-sized houses of 200 square meters in 24 h. This project was achieved at the cost of $4,800. The process of 3D printing involves the use of CAD for the architectural work plan, a layer by layer mix of high-quality cement and glass fibre material. The houses were constructed from recycled industrial and construction waste materials, including concrete, cement, rubble, steel, and binders. The 3D houses were printed in prefabricated panels which the 3D printer invented by WinSun produced. These panels, which were printed in a central factory, were assembled and installed on the site. The resultant benefits of 3D printing in the actualization of the project include low cost, speed, less labour requirement and solution to the housing crises confronting a massive population in China.

6.4 Benefits of 3D Printing Addictive Manufacturing in Construction

The introduction of 3D Printing Technology/Addictive Manufacturing offers a solution to some of the problems faced in the construction industry. As a result, contractors have discovered the adoption of 3D has a way to gain a competitive advantage in the industry. These benefits are discussed below:

(1) **Speedy Delivery of Construction Project:** 3D Printing is considered one of the emerging technologies of construction 4.0, which has greatly reduced construction time compared to conventional construction methods. Ordinarily, a project that might take weeks or months to complete can now be accomplished within hours or days by adopting 3D printing. This is possible as the technology is fully automated and operates faster with minimal human error than the traditional construction method. This benefit of speed production allows contractors to execute projects on time and manoeuvre other projects.

(2) **Cost-Effectiveness:** Cost reduction is a significant factor that justifies the adoption of new technology. The use of 3D printing technology in the construction process is an important milestone that has been achieved. This is made possible through the reduction in the process of construction, reduction in labour cost due to automation, reduction in energy cost due to lightweight end-product and non-requirement of expensive tools such as moulds and patterns. Also, travelling costs are reduced because 3D can create and complete a product without moving from one place to the other. Also, cost-effectiveness through the adoption of 3D printing in construction can be achieved by using recycled materials, which reduces material wastages, transportation and storage. Consequently, when comparing addictive manufacturing and traditional construction, these factors have radically reduced production costs.

(3) **Waste Reduction and Sustainability:** 3D printing has proved to be a double-edged sword for waste management, and the traditional construction process is well known for waste. This includes broken and damaged materials such as bricks, wood, and concrete, often accumulated during construction works. However, 3D printing is designed to use the exact materials specifically required to build products with layer-to-layer processes. Also, the 3D technology can recycle industrial and construction waste and convert it to building materials. As a result, waste is reduced to the barest limit. In terms of sustainability, 3D technology releases less carbon dioxide, which can resolve the challenge of large emissions in the construction industry.

(4) **Innovative Design Solutions:** One of the added advantages of 3D printing is the innovative solutions it offers to the construction industry. This technology avails the building designers, such as architects, to construct complex designs that traditional methods cannot achieve in terms of labour and cost. Also, 3D printing presents several designs and shapes, such as unique facades and curved walls, unlike the rectangular shape structures, which is familiar to the traditional construction method. The possibility of innovative building designs improves clients' satisfaction as 3D printing gives them the opportunity to personalized their design structure. The creation of complex structures using 3D does not come with a higher price than conventional methods in which price increases with complexity.

(5) **Improve Safety in Construction the Industry:** Construction is well known as one of the most dangerous industrialise with a high recorded injury and accident record. 3D printing offers a solution to the challenge of safety in the construction industry. This technology which is considered construction 4.0

technology, has automated the construction process, thereby reducing labour, risks and accidents among construction workers. 3D printing is designed to perform most of the tedious work carried out on the construction site. As a result, this has helped minimize the rate of injuries and accidents on the construction site. Furthermore, 3D technology can detect on time risks that might endanger the workers during the construction process. Consequently, project managers and their teams become proactive in resolving such issues to promote safety.

(6) **Better Durability of Building Project:** The durability of a building is of great importance in the construction industry. 3D printing is well known for designing structures that can withstand adverse environmental conditions such as earthquakes and hurricanes. A recent discovery has shown that the method by which materials are created and assembled using 3D printers contributes to structural elements' durability. The advantage of securing building structures that will last for a long time goes a long way to improved clients' satisfaction, reduces the incidence of repairs, and increases the construction industry's productivity. Examples

(7) **Improved Project Management:** The construction industry is reaping a lot of benefits with 3D printing technology. This has brought about a great improvement in project planning, execution and timely delivery. This is made possible as 3D printing helps resolve several challenges of the traditional construction methods. Hence, with the adoption of 3D printing technology, the benefits of better and more accurate designs, reduction in safety risk, accurate cost projection, lesser labour, reduction of materials, etc., is achieved.

6.5 Risks Associated with 3D Printing in Construction

The introduction of new technology such as 3D printing presents new risks. Despite the benefits and potentials of 3D printing technology in the construction sector, some risks come with its implementation. In light of this, it becomes crucial that these risks associated with 3D printing are addressed to mitigate the adverse effect which might hinder its effectiveness. Below are some of the associated risks.

6.5.1 Health Risk

3D printing presents health risks in terms of worker's exposure to the emission from 3D printers. Some of the materials such as nylon, thermoplastic, lay wood and lay brick used for 3D printing have a level of toxicity that can be harmful to human health. Also, the generation of ultrafine particles, explosive organic, polymers and heated filaments released by most 3D printers is hazardous to human health. This can lead to health challenges such as heart disease, asthma, and stroke. Some of the ways of minimizing these health risks are by designing 3D printers that limit the emission of particles, the use of materials with low emitting filament and ensuring the 3D printers are operated in a well-ventilated environment,

6.5.2 Cyber Risk and Intellectual Property Risk

With the increase in internet use and the digitalization of many operations, digital designs and infringement on patents have become rampant. The ability of the 3D printer to design and print a product at any place, any time through digital means presents cyber and intellectual property risk. 3D printing makes use of Computer-aided drafting (CAD), which produces files that contain patented information. Loss of these files can lead to theft of designs. Also, there has been a rise in hackers who can gain access to information stored digitally, and this presents another risk to the use of 3D printing being a data-based technology. Furthermore, the danger of a compromised supply chain may arise as manufacturers face the risk of producing a product with unsafe 3D printed parts, which might cause damage.

6.5.3 Legal and Regulation Risk

The adoption of 3D printing in construction presents new legal risks and implications. First and foremost, the regulations of 3D printing have not been fully implemented in the construction industry compared to other sectors. Also, a defect or improper functioning of 3D technology can cause damage to property and humans, thereby incurring liability on the key players connected to the manufacturing of the technology and project management. Consequently, this might lead to a lawsuit. Similarly, in terms of regulations, there is the risk of government regulations such as inspection requirements. The scheduling procedures of the project may conflict with the increased speed of the 3D printer in completing a project. As a result, government agencies may be required to adjust to new technologies to meet the demand of 3D technology, which might be challenging to attain in struggling economies (developing and under-develop economies). Currently, there's much uncertainty in this aspect of 3D printing in construction. 3D printing will unlikely make too much of a mark in the construction sector until laws and regulations are clearly defined. (Ellis, 2018).

6.5.4 Risk of Job Loss

The construction industry, which is labour intensive, faces a threat of labour reduction, ultimately leading to job loss due to adopting 3D printing. 3D printing is a wholly digitalized process of designing and manufacturing products, thereby making construction simple and easy. Products are designed and printed from the beginning to the end with few operators. This technology is currently applied to almost all the construction processes, and as a result, the demand for labour is greatly reduced. Although it is considered a viable tool that can resolve the problem of shortage of labour, there is a greater risk of job loss, especially among unskilled labour. Consequently, the increase in the use of 3D printers in the construction industry will translate into lesser workers on construction sites, as most tasks become automated.

6.5.5 Management Risk

The newness of 3D printing in construction presents some managerial threats. These include a shortage of experts in 3D printing technology, as 3D printing requires a more specific skill set that would have to pull from a slimmer and more niche group of candidates. Construction labour shortages are already a problem, and finding qualified workers to employ in 3D printing construction environments could prove to be even more challenging. Also, a 3D printer presents construction risks that might surface due to errors in the digital model, which create some problems for the site resulting from reworking. Furthermore, managing quality is another factor in 3D printing in the construction industry; being an automated process, there is a need for constant monitoring of the process by real humans. Otherwise, the process could be an expensive mess (Ellis, 2018).

6.5.6 Material Shortage Risk

The materials for 3D printing are crucial to the smooth running of the technology. The risks associated with 3D printing materials arise from limited availability of materials or incompatibility of materials with the technology. When these materials are not available, it can paralyse the operation of the technology. Furthermore, the use of materials not compatible with the specification of 3D technology presents a great risk of liability for defects on the project contractors.

6.5.7 Financial Risk

3D printing technology requires a high cost of implementation. Perhaps the biggest challenge to the widespread adoption of 3D printing technology on construction sites is the high cost of purchasing or renting such equipment and the logistics involved in getting these large 3D printers to the worksite. 3D printers are costly, and that upfront purchase cost doesn't include materials or maintenance. Due to the high cost of setting up 3D printers for construction work, there is a low demand for 3D technology.

6.6 Skills Requirement for 3D Printing Application in Construction

The 3D printing/addictive manufacturing technology is gaining more grounds and transforming project management in the construction industry. Accordingly, skilled personnel with certain skills are required for the effective implementation of this technology. These skills are needed to comprehend digital and masonry technology to detect glitches in the digital model during construction (Brehm, 2019; Hossain et al., 2020). Consequently, training is required for current workers to adapt to the new working processes. Below are some relevant skills for the effective application of 3D printing in construction.

6.6.1 Technical Skill

Technical skill is vital in understanding and utilizing the 3D designs software and tools. A major application of 3D printing is prototyping which involves an experimental and developmental process of designing a product from a prototype. Technical skills are required all through the 3D printing process to ensure that the final products printed are in line with the prototype. This involves inputting relevant data in an orderly and pre-planned way into the 3D printing machine, producing the specific output. Technical skill is essential because any error in the printing process is of great consequence to the product outcome.

6.6.2 Software Designing Skill

The design of any project is key to the successful accomplishment of the desired outcome. 3D printing involves creating and designing new object from scratch before it is transformed to 3D model for printing. The 3D process starts with the building of a prototype using digital designs using a computer. These designs are produced with the software. More often, the 3D process involves using computer-aided design (CAD) software as it gives the detailed information required to build a product. Construction professionals such as architects, engineers, and project managers use CAD to create precise designs and drawings of proposed building projects. Consequently, the increasing application of 3D in the construction industry demands advanced software design skills for effective implementation.

6.6.3 Modelling Skill

The arrival of the 3D models has transformed Architecture Engineering and Construction (AEC). 3D models offer better architectural designs, diagrams, and building plans—this avails the architect's opportunity to communicate effectively and provide better and satisfactory service to their clients. However, the procedure of 3D printing technology lies in the ability of the operator to conceptualize the desired object and precisely create a digital representation. As a result, the modelling skill with emphasis on CAD is required to construct models for 3D printing. CAD modelling is a complex process that entails learning software, tools, basic commands, visualising the conceived ideas in 3D, and converting the model into a printable 3D object. Developing the modelling skill becomes essential for a construction professional who doesn't want to be limited to basic designs and models.

6.6.4 Creativity Skill

Creativity is a required skill for the effective utilization of 3D printing technology. Every project begins with conceiving an idea that can be brought to reality through 3D technology. This technology is a digital approach that allows for innovative designs and artistic creativity such as drawing, moulding and painting. All that is required to deliver a product with the technology is a blueprint. Consequently, 3D printing involves the process of creative thinking outside the box to generate designs and objects that might be difficult or impossible with traditional construction methods. Accordingly, professionals in the construction industry, especially architects and designers, need to develop their creative skills to explore the full benefits of 3D printing.

6.7 Summary

3D or Addictive Manufacturing has become an indispensable technology for the construction industry. The application of 3D printing in civil engineering is promising as it is particularly helpful in creating buildings with complicated shapes. The construction industry has been slow in adopting 3D printing technology. However, since the 3D printing technology has just started, it is believed that many more developments are still uprising, and some have been field-tested, some used for stereotyping etc. The current advancement in technology has shown diverse advantages in the construction industry. Some of the benefits include low cost, higher accuracy (resulting from eliminating human error due to automation), material saving, reduction in labour, increased work safety, time-saving, and design of complex 3D shape composition of pillars. 3D technology has proved to be worthwhile, and further advancement would benefit the construction industry. Also, the implementation of 3D printing presents some threats. These include health, legal regulations, loss of job, management, material shortage, financial, cyber and intellectual property risk. For the construction industry to fully implement 3D printing, there is a need for reskilling and upskilling. Effective adoption of this technology requires both soft and technical skills. These include software designing, modelling, graphics, safety management, creativity, critical thinking, problem-solving, finishing and intellectual property protection.

References

Alberta, A., & Paolo, F. (2020). Special issue on materials development by additive manufacturing techniques. *Applied Science, 10*(5119) 1–4. doi:https://doi.org/10.3390/app10155119.

Brehm, E. (2019). Robots for masonry construction—Status quo and thoughts for the German market. *Mauerwerk, 23*, 87–94. https://doi.org/10.1002/dama.201900004.

Buchanan, C., & Gardner, L. (2019). Metal 3D printing in construction: A review of methods, research, applications, opportunities and challenges. *Engineering Structure, 180*, 332–348.

Cherdo, L. (2020). The 13 best construction 3D printers in 2020. Retrieved from ANIWAA. https://www.aniwaa.com/buyers-guide/3d-printers/house-3d-printer-construction/.

Ellis, G. (2018). Retrieved from the truth about 3D printing in construction. https://blog.plangrid.com/2018/11/the-truth-about-3d-printing-in-construction/.

GaudilliÈre, J., Duballet, R., Bouyssou C., Mallet, A., Roux, P., Zakeri, M., & Dirrenberger, J. (2018). Large-scale additive manufacturing of ultra-high-performance concrete of integrated formwork for truss-shaped pillars. Robotic Fabrication in Architecture, Art and Design 2018, *Springer International Publishing*, pp. 459–472, doi:https://doi.org/10.1007/978-3-319-92294-2_35. hal-01904661

Giacomelli, P., & Smedberg, A. (2014). The eve of 3D printing in telemedicine: state of the art and future challenges.

Hambach, M., & Volkmer, D. (2017). Properties of 3D-printed fiber-reinforced portland cement paste. *Cement Concr Compos, 79*, 62–70. http://refhub.elsevier.com/S1359-8368(17)34294-4/sref144.

Hamad, A. l., Fady, A., & Abdel-Hamid, M. (2018). Large-Scale 3D Printing: The Way Forward. *IOP Conference Series: Materials Science and Engineering*. doi:https://doi.org/10.1088/1757-899X/324/1/012088.

Herzog, D., Seyda, V., Wycisk, E., & Emmelmann, C. (2016). Additive manufacturing of metals. *Acta Mater, 117*, 371–92. http://refhub.elsevier.com/S1359-8368(17)34294-4/sref40.

Hossain, A., Zhumabekova, A., Suvash, C. P., & Jong, R. K. (2020). A Review of 3D Printing in Construction and its Impact on the Labour Market Md. *Sustainability, 12, 8492*, 1–21. doi:https://doi.org/10.3390/su12208492.

Izabela, H., Anna, G., & Roman, P. (2016). 3D printing of buildings and building components as the future of sustainable construction? *Procedia Engineering*, 292–299.

Khoshnevis, B. (2004). Automated construction by contour crafting—related robotics and information technologies. *Autom ConStruct, 13*(1), 5–19. http://refhub.elsevier.com/S1359-8368(17)34294-4/sref19.

Mehmet, S., & Yusuf, C. K. (2017). 3D printing of buildings: construction of the sustainable houses of the future by BIM. In *9th International Conference on Sustainability in Energy and Buildings, Energy Procedia 134,* (pp. 702–711).

Ngo, T. D., Alireza, K., Gabriele, I., Nguyena K. T., & David, H. (2018). Additive manufacturing (3D printing): a review of materials, methods, applications and challenges. *Composites Part B 143* 172–196. https://doi.org/10.1016/j.compositesb.2018.02.012.

Parandoush, P. & Lin, D. (2017). A review on additive manufacturing of polymer-fiber composites. *Compos. Struct., 182,* 36–53. https://doi.org/10.1016/j.compstruct.2017.08.088.

Raspalli, F., & Banon, C. (2017). 3D printing architecture towards functional space. In *Conference Proceedings of Learning, Adapting and Prototyping—Proceedings of the 23rd CAADRIA Conference, Beijing, China,* (pp. 215–224).

Sagheer, R., Ambarish, K & Jay, S. (2018). 3D construction printing—a review with contemporary method of decarbonisation and cost-benefit analysis. In *First International Conference on 3D Construction Printing (3DcP), and the 6th International Conference on Innovative Production and Construction (IPC 2018),* Melbourne, Australia.

Shakir, Q. M. (2019). 3D printing of houses. *Retrieved from Research Gate.* https://doi.org/10.13140/RG.2.2.29453.08168.

Travitzky, N., Bonet, A., Dermeik, B., Fey, T., Filbert-Demut, I., Schlier, L., Schlordt, T., & Greil, P. (2014). Additive manufacturing of ceramic-based materials. *Advanced Engineering Materials, 16*(6), 729–54. http://refhub.elsevier.com/S1359-8368(17)34294-4/sref18.

Wang, J.C., Dommati, H., & Hsieh, S. J. (2019). Review of additive manufacturing methods for high-performance ceramic materials. *International Journal of Advanced Manufacturing Technology, 103,* 2627–2647. https://doi.org/10.1007/s00170-019-03669-3.

Withell, A., Diegel, O., Grupp, I., Reay, S., De Beer, D., Potgieter, J. (2012) Porous ceramic filters through 3D printing. In *Innovative developments in virtual and physical prototyping—proceedings of the 5th international conference on advanced research and rapid prototyping.* (pp. 313–318). http://refhub.elsevier.com/S1359-8368(17)34294-4/sref127.

Chapter 7
Robotic Construction Technology

7.1 Introduction

Robotics, also known as automation, have contributed to improving the construction process in all aspects. Over the years, the construction industry has been with an image of hard work of long hours in a hazardous working environment. However, the introduction of automation and robotics technology has brought about a solution to this problem. Robotics and automation often go hand-in-hand. Historically, the introduction of the computer brought about programmable automation in which machines are controlled by fully integrated computer systems and are flexible enough to be reprogrammed according to a specific task. Automation is a broad terminology that increases productivity and efficiency, involving less effort. Automation such as robotics are machines attributed with high-level programming and machinery abilities capable of making decisions using sensors, recognizing errors and faults to generate solutions to those faults (Bolivar, 2019). A robot is a machine, while robotics is the application of robots. Robotics is defined as "the science of designing, building, and applying robots." (Bolivar, 2019).

Robots are equipment built on an automated system to carry out a specific related task. A robot is a reprogrammable, multifunctional manipulator designed to: move material, parts, tools, or specialized devices through variable programmed motions for various tasks' performance (Bolivar, 2019). Robotics comprises the ability of tools with multi-axis flexibility to perform diversified tasks by themselves. This allows the mass- customisation concept to be applied in the production. Robotics construction can be traced to the 1980s by introducing single-purpose robots designed to perform specific jobs such as concrete finishing or welding and majorly dangerous, difficult, and dirty activities. Afterwards, several large construction companies in several countries started to commit considerable investments in robotics technologies to solve the high cost of projects and cost overrun. In the construction sector, robots were first developed in Japan and applied to off-site prefabrication of building to increase the quality of building components for modular homes (Bock et al., 2012). In the last

O. Adepoju et al., *Re-skilling Human Resources for Construction 4.0*, Springer Tracts in Civil Engineering, https://doi.org/10.1007/978-3-030-85973-2_7

decades, construction robotics have been generating much interest in the construction community. Robotic systems have become common in many manufacturing and production operations because they have proven to be more robust, safe, efficient, accurate and productive.

In the construction industry, robot technology is increasingly being adopted to build construction sites, efficiently reduce construction time, improve working conditions, and increase safety by replacing humans in dangerous operations.

Automation and robotics in construction became necessary to resolve some of the problems associated with the construction industry, such as lack of skilled personnel, decreasing quality and productivity, occupational safety and substandard working condition (Mahbub, 2012). Robotics in construction involves the incorporation of mechanical technology such as machines, engines with computer technology (automated systems, programming languages, control system, etc.) to achieve greater productivity and ensure safety during construction activities such as earth-working, data processing, monitoring and feedback, heaving lifting of equipment etc. (Son et al., 2010; Yahya et al., 2019). Modern robots are usually re-programmable, having multifunctional manipulators designed to carry materials, tools, parts, or other devices through variable programmed motions for the performance of series of tasks occurring in the construction process (Krom, 1997; Stein et al., 2002; Yahya et al., 2019). The construction robots require engines, batteries or motors and are able to drive themselves. The development of robots in the construction industry is gradually moving from factories to actual construction sites.

In construction, robotics technologies can be implemented in all stages of the construction life-cycle, including the design, construction of the building on-site, maintenance after completion of a building, and demolition of the building (Struková & Líška, 2012). These technologies can also perform functions such as planning, cost estimation, construction schedules and project management. One major setback of robots and automation is that they tend to operate within a given set of programming. According to Bolivar (2019), tasks performed by robots are programmed according to certain conditions and factors, making it impossible for robots to deviate or change from the planned task or even plan new things unless they are reprogrammed. Recent technological development is moving to fix these issues with the help of artificial intelligence and machine learning. Recently, the use of robots in construction has grown at a tremendous level, and their adoption on construction sites include positioning, inlaying, attaching, coating, concreting, connecting, building, covering, jointing, scaffolding, demolishing, finishing, tunnelling, repairing elements and inspecting (Bolivia, 2019).

7.1.1 Components of a Robot

According to Anudari (2019), a robotic system comprised the following parts:

(i) **Body**: The body parts have similar functions to the human body parts. It serves as the body to other internal parts of a robot, and it encloses the internal components giving it a rigid shape and protecting them from external damage. Also, the body provides a structure where other essential parts like sensors can be attached. Robots body parts are built to be strong and vibration resistant.

(ii) **Locomotion**: Fully automated or not, robots have to move. The locomotion part makes sure the robot can move from one point or place to another on its own without being physically carried about by a human. Robots locomotive equipment may be in the form of legs (can walk), Wheels (can drive itself), wings (can fly).

(iii) **Sensors**: For identifying the environment, sensors are necessary for a robot because they receive information from the environment and send electric signals to the control unit. Sensors are used to scan the environment to avoid robots crashing into some objects and mapping task environments.

(iv) **Manipulators**: These are used for movement and navigation during work-in-progress. The type and purpose of the robots will determine the number of manipulators such robots would have. Manipulators differ in shape and size. They serve as the arm for a robot when interacting with the work environment. The simplest manipulators have three Degrees of Freedom (DOF), which describes the number of the axis that it can move. Robots with three DOF can move in x-axis up and down, in y-axis back and forth and in z-axis rotate. The higher DOF of a manipulator, the more complex is the robot.

(v) **End Effectors**: these are tools surrounding the robot's manipulator and are used for grabbing, screwing, holding, etc. Depending on the purpose of the robots, the end effectors can be grippers, hammers, spikes or even screwdrivers.

(vi) **Control System**: In general, control systems are known for data processing. The control system is the robot's brain, where all information is received, processed and transmitted to various parts where they are needed. A Control system is a combination of hardware and software programming which requires coding. Artificial Intelligence should increase the functional ability of the control system, helping the robots make decisions and learn from previous experience based on received data.

(vii) **Power Supply**: Robots need the energy to work and function properly. In the construction world, where robots are applied onsite, such robots need replaceable or rechargeable batteries to work correctly.

7.1.2 Types of Robotics Construction

Robots in construction can operate either on the ground or in the air. However, there are two major types of robots in construction: ground robots and aerial robots. Ground robots are robots that move on the ground manually or automatically using a ground mechanism such as wheels. Theoretically, there is no ad-hoc consensus regarding the categorization of construction robotics. The lines between categories are constantly moved or blurred by new developments in technologies. The difference in categorization is based on individual authors perception and understanding of robotics in construction. Oesterreich and Teuteberg (2016) also listed several types of robotics technologies used in the construction industry as:

- Reality computing technology (Laser Scanners, Rovers, Unmanned Aerial Systems)
- Simulation technology (Augmented Reality, Mixed Reality, Virtual Reality)
- Automation technology (Additive Manufacturing, Autonomous Vehicles, Prefabrication and Modularization Robotics systems)
- Smart Technology (Artificial Intelligence, Big Data, Cloud computing, Cyber-Physical Systems, Human–Computer, Interaction Internet of Things (IoT), Radio-Frequency Identification)

Li (2018) categorized robots in the construction industry into three: traditional, wearable robotics, and arms robotics.

(i) **Traditional robot** : These types of robots are controlled by computers or other kinds of stimuli. They are designed to construct the superstructure of buildings autonomously and are applied to different construction sections. These include, e.g. climbing robots for skyscrapers, traffic marshal robots to direct cars during bridge construction and high maintenance. Traditional robots are also deployed to reduced onsite work hazards associated with the installation of heavy building materials by manual labours (Li, 2018; Li & Ng, 2017)

(ii) **Wearable robotic/exoskeleton**: This is a support robot used to support work stress on the human body. These types of robots come in the form of a mechanical suit to be worn by a human controller, which assists to lift heavy materials such as reinforcing bars, wall panels etc., which ordinarily cannot be lifted by an individual. These robots are designed to fix a fundamental weakness of humans, which diminishes strength and fatigue. On the other hand, the exoskeleton is designed to move naturally with the wearer and are usually adjustable to suit different body types. Research prediction reveals that there will likely be an increase in demand for wearable robots due to the shortage and ageing of construction workers (Li, 2018; Li & Ng, 2017).

(iii) **Robotic arms**: These are robots designed to imitate human arms and are usually lightweight. These types of robots are small but have an excellent capacity for carrying out all types of work conducted by robots. They also serve as a tool for lifting and placing objects the same way a human hand would have done (Fig. 7.1).

Delgado et al. (2019) categorized robotics alongside a system of operation on-site automated, off-site prefabrication systems, robotic systems, drones, autonomous vehicles and exoskeletons.

(i) **Off-site automated prefabrication systems**: Automated robotic systems require no human effort to operate. These offsite facilities build, recycle, and transform construction materials such as concrete, steel, wood, bricks, etc. to high-level prefabricated construction components (concrete prefab elements, steel trusses, wood structural elements, etc.). Their operations improve the quality of prefabrication components. Examples include large-scale prefabrication (LSP) with high-level building components and 3D printing, also known as the additive manufacturing technique.

Fig. 7.1 Exoskeleton robot (Ruggiero et al., 2016)

(ii) **On-site automated and robotic systems**: This category involves robotic systems directly applied on the construction site. At first, this category of robotics were single-task construction robots (STCRs), which can only carry out a singular specific task in a repetitive manner. An example is a robotic arm that can be used to carry out simple tasks such as bricklaying, wall painting, scaffolding, welding, concrete spraying etc. This robotic system is very flexible because it could be easily adapted to other traditional construction methods. However, this approach generates other challenges such as the need for additional health & safety requirements, the difficulty of parallelising and integrating with human worker activities, and the lack of integration with downstream and upstream activities.

(iii) **Drones and autonomous vehicles**: This category of robotic construction comprises terrestrial, aerial or nautical automobiles that steered remotely. Drones and autonomous vehicles can be used for various construction tasks, including surveying and monitoring, accessing dangerous environments to protect workers from high-risk areas and excavating, demolition and transporting materials (Delgado et al., 2019). Terrestrial vehicles are often used for groundwork construction such as building foundations, rehabilitation of road networks, monitoring and inspecting ongoing projects and handling heavy material onsite. Aerial vehicles are mainly used for surveying, inspection and monitoring.

(iv) **Exoskeletons**: Exoskeletons are complementary robots used to augment human strength and capacity. They are wearable devices that are integrated to work together with the user, which performs the task instead of a fully robotic system that autonomously carries out tasks without human interference. Exoskeletons can help to minimize fatigue, injuries, disabilities and stress accruing to the repetitive and heavy nature of daily tasks among construction workers. This robotic has helped to proffer solutions to the ageing problem in construction industry manpower as it could perform dangerous and challenging tasks stress-free (Delgado et al., 2019).

7.2 Application of Robotic Technology in Construction

In the modern age, the application of robots in construction activities has increased. Robots are applied to construction tasks to gain speed, accuracy, safety and quality project. Robotics has been used for several sections of construction, both on-site and off-site. Below are some of the areas where robots have been used in construction. From the work of Ruggiero et al. (2016) and Bolivar (2019). The technological milestones that have been reached in the construction industry have been made possible by the use of robots in performing the following functions:

7.2.1 Painting and Spraying

Painting is a significant activity in building construction that is time-consuming and labour intensive. It is carried out to beautify and protect the surface of a building from weather conditions and decay. Robots have been designed to apply paint/spray paint on both the interior and exterior surfaces of the constructed buildings. These robots can accomplish the painting task faster the humans. The application of robotics to painting promotes the accuracy of painting quality (smooth, evenly sprayed surface with uniform spray quality) across all surface areas. This technology has helped to address some of the challenges and risks of construction. Some of these challenges and risks include labour shortage, painting hazards, high rise buildings and inaccuracy, and variation in painting quality by different professionals. Painting robots has helped to solve labour shortage as one person is required to operate the technology. This has enhanced the production rate about five times the efficiency of conventional methods regarding speed, waste, accuracy etc. (Akshatha et al., 2017). Also, the risk of injury or death that painters face while working at height is minimized.

Consequently, there is the speed of job accomplishment and increase productivity in the construction industry. The painting robot is designed using few sheets of steel, a conveyor shaft, a spray gun and a controller unit to control the entire operation of the robot. This robot is simple to install, durable, efficient, and has simple control systems with control noise vibration, thereby operating without noise and vibration (Keerthanaa et al., 2013) (Fig. 7.2). For effective performance, a painting robot requires three (3) degrees of freedom; X (to approach the wall), Y (to move between successive strips) and Z (to adjust the orientation facing the wall). Figure 7.3a and b depict a typical example of a painting robot.

Fig. 7.2 Risks painter faces during painting (Jayaraj & Divakar, 2018)

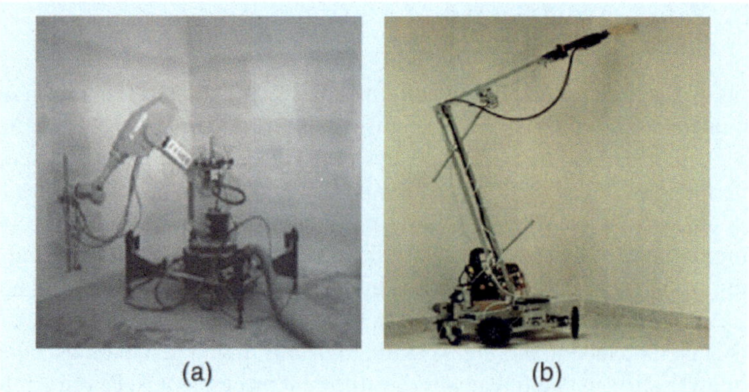

Fig. 7.3 Painting robots (Ahsan et al., 2018)

7.2.2 *Demolition*

Demolition is an essential aspect of construction, most specifically during the renovation and reconstruction of existing buildings. This activity is conducted either for the purpose of redesigning, restructuring and renovation. The introduction of this activity has resolved the challenges of hazard and time consumption of demolition. The prime benefit of using robotics for the demolition process is that it promotes safety, increased efficiency and effectiveness. It allows the operator to operate the machine by staying at a safe distance from the contaminants and debris, making them safer than handheld machines (Ruggiero et al., 2016). Also, it is used in the restricted space and selective demolition works (Fig. 7.4). The precise control enables the demolition of needed sections while leaving the remaining sections unscattered (Akshatha et al., 2017) (Fig. 7.5).

Fig. 7.4 Demolition robot (Akshatha et al., 2017)

Fig. 7.5 Demolition machines (Cieslaki et al., 2015)

Currently, there are three types of demolition robots which are:

1. **Multi-tooled**: A multi-tooled robot is designed with different kinds of tools on the robot's arm. This comprised multiple tools placed at the end of a robotic arm on the demolition robot.
2. **Hydro-powered**: These robots use hydro generated pressure to break down the materials such as walls, weak concrete and beam with ease. As a result, it prevents the chances of air pollution and contamination.
3. **Eco-friendly**: Eco-friendly demolition robot absorbs the debris and material removed (without environmental contamination) in the process of demolition for recycling if necessary (Figs. 7.6, 7.7 and 7.8).

7.2.3 Bricklaying and Plastering

Bricklaying is considered one of the most tedious and monotonous in the construction process. It entails standing, lifting and kneeling and working on high rise buildings. As a result of the tedious nature of this task, there is a probability increase in risk. The advent of the robot has helped to resolve the risks associated with bricklaying. In recent times, bricklaying robots are used in construction sites to execute such monotonous tasks efficiently. Bricklaying robot is not a recent technology. The first bricklaying device named Motor Mason was patented in 1904. This

Fig. 7.6 Multi tooled demolition robot (Ruggiero et al., 2016)

Fig. 7.7 Hydro-powered demolition robot (Ruggiero et al., 2016)

Fig. 7.8 Eco-friendly demolition robot (Ruggiero et al., 2016)

device is a mechanical robot that required three people who are skilled in brick-laying for operation. This technology operates five to 10 times faster than humans. However, this technology had major drawbacks like regular breakdown and large size restrict its placement around the corners and complex structures like windows. Recent bricklaying robots have added advantages such as: it utilizes arms of industrial robots to perform repetitive tasks and can assemble the masonry structure of the building whereby human-operated robot mixes bonding agent or cement and makes it possible to create structures. This ordinarily cannot be created by traditional methods. Example of bricklaying robot which has been tested and applied include SAM 100 and Hadrian.

Hadrian is a more static robot attached to a position at the start to the end of its bricklaying task with only the manipulators moving. On the other hand, SAM 100 (Fig. 7.9), a complete brickwork robot realized, has its manipulators mounted on a movable chassis. The manipulator is supplied with bricks by a device then further moves it around the wall. However, one major bottleneck of applying robotics to the bricklaying section is positioning, i.e. creating an effective algorithm to move the different sites with different characteristics. Nevertheless, algorithms like Free Movement Algorithm (FMA) and Brickwork Passing Algorithm were developed to tackle the movement issue (Malakhov et al., 2020). The recent design called Semi-Automated Mason (SAM), designed by New York-based Construction Robotics can

Fig. 7.9 SAM 100 (Dakhi & Lafhaj, 2017)

lay 3000 bricks in an eight-hour shift, which is more than a traditional mason. Consequently, the main benefit of this technology is efficiency and time reduction. The only limitation to this technology is the high cost of the installation (Fig. 7.10).

Fig. 7.10 Bricklaying robot (Semra & Hafize, 2017)

7.2.4 Construction Welding

Welding is an essential and instrumental activity towards the success of any construction project. It connects metals in a process called fusion which entails heating, intermixing and cooling. Welding robots or robotic welding is an automated technology designed to join anything steel (beams, columns, riggings used for concrete slabs etc.) together. These robots can perform accurate welding and manoeuvre easily in difficult locations. In the construction business, such technology is widely used in building skyscrapers by wielding the iron beam used in the structure. Also, it is used in welding the rings of the concrete slabs for large-scale projects. This technology is most widely used in Japan, with minimal application in other countries. Adopting welding robots in construction presents a lot of advantages. It is faster, operates without interruption, has better weld quality, promotes a safe work environment, conserves materials, and cuts down labour costs. It increases the efficiency and productivity of the construction industry. The shortcoming to this technology is that it is a stationary device that needs to be installed in the desired work location (Fig. 7.11).

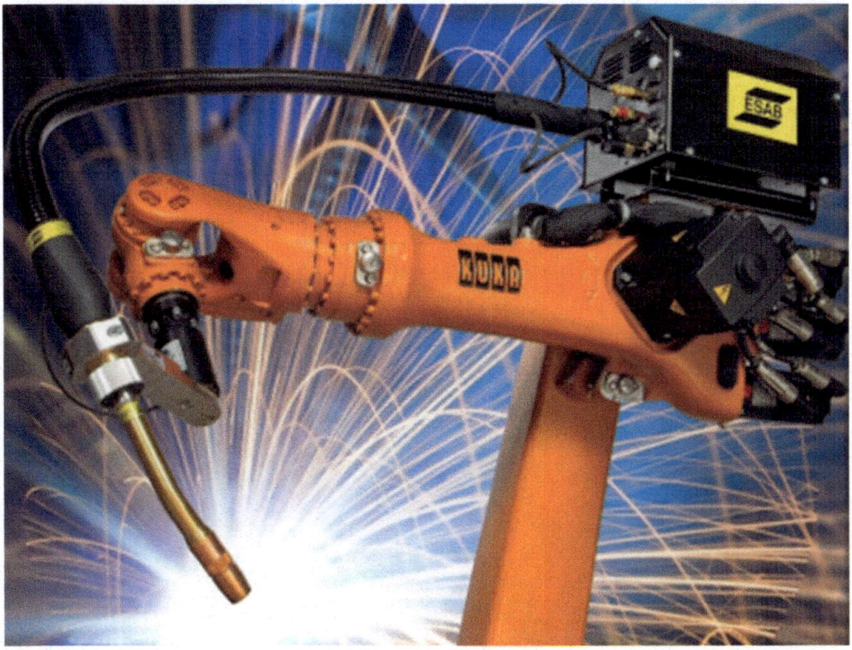

Fig. 7.11 Welding robot (Ruggiero et al., 2016)

7.2.5 *Transportation in Construction*

Transportation is a vital function in construction. It ensures that building materials are conveyed from the manufacturers or where it has been procured to the construction site. The movement of materials is an essential part of the construction process because all equipment and materials have to be moved around during different tasks, which may be burdensome. These movements by humans tend to lead to stress and body pain which is the major problem transportation robots are designed to fix. Recent development has shown the application of robots for transportation on construction site. Hence the robots are used for heavy lifting while humans do the fixing. This important activity leverage by the robots ensures prompt delivery of the project and enhances the supply chain towards achieving a competitive advantage. By conceptual application, almost all robots (except static) used on-site are mounted on transportation machines or robots for movement. Robots transportation can, therefore, be categorised into two: aerial and ground transportation robots. As explained in the previous chapters, Drone and Forklift robots are majorly used to transport light, heavy, small or large-sized materials on-site. Technology such as drones and autonomous vehicles are helping to get the work done faster and minimizing the hazard-prone to the manual worker. Transportation drones are majorly used to carry lightweight construction materials to areas needed. This technology can easily access high locations in less time and without stress than human workers who can get tired and stressed up (Fig. 7.12). The downside of such technology is that due to variation in the wind, efficiency and accuracy may be reduced. The forklift robots (Fig. 7.13) also work on the ground and can be used to transport large and heavy materials from place to place. These robots are designed for reducing the human effort used in

Fig. 7.12 Transportation drone (Ruggiero, Salvo, & Laurent, 2016)

Fig. 7.13 Forklift robot (Bolivar, 2019)

carrying heavy objects to any construction site. For the effective operation of forklift robots, there is the need to have a flat ground of smooth operation, which is almost impossible to get on a construction site. A road network always has to be created for proper application with the construction topography that'll be flat for easy passage of a forklift robot. Another critical consideration is that the construction site's terrain network map has been programmed into the robot.

7.2.6 Surveying and Monitoring

Construction robotics such as drones is used for surveying and monitoring during construction project execution. By navigation of the construction site, these drones help scan, assess and provide vital information about the progress of a construction project. The information collected by the drones promotes safety by detecting potential problems and errors early. These types of drones are equipped with high-resolution cameras for aerial shots and video recording. This type of drone was applied to the Mini Sky City in Japan, where all the progress of the construction was monitored from the foundation level up to the finishing. In recent times, drones with artificial intelligence can help to predict tasks required ahead (Fig. 7.14).

Fig. 7.14 Surveillance and monitoring drone (Ruggiero et al., 2016)

7.2.7 Roadwork

Road and highways construction is very significant to the socio-economic development of a nation. It promotes effective and efficient transportation of people and goods from one location to the other. Roadwork construction involves several tedious tasks such as cut and fill, grading, base preparation and placement, surface material placement, curbing and guardrail placement and road maintenance. The introduction of robotic construction technologies is helping to resolve problems causing disruption and congestion for travellers on the roads. This type of robot is currently in its developmental process and testing state, which are majorly developed primarily for revamping and repainting roadways (roads and bridges). Currently, large machinery is required to repaint and fix small potholes and other cracks inroads. This technology is smaller in scale and can do smaller patch jobs on potholes and cracks. The repainting robots allow for the same type of small job scale fix jobs. Their main abilities are that they alleviate the need for a large workforce and machinery for relatively small jobs (Ruggiero et al., 2016). Ultimately, robotic roadwork construction offers several advantages, including speed up construction time, facilitating problem reporting, better traffic control, and reducing disruptive excavation during road construction (Fig. 7.15).

7.2.8 Interior Decoration

Interior decoration is an important process in completing a building project, and this process includes wiring, plumbing plastering, masonry, painting, and air conditioning. Interior decoration is time consuming, monotonous, draining, hazardous and demands a high level of precision. As a result, modern technology such as robots presents excellent opportunities for accurate, speedy, and easy interior decoration (Fig. 7.16). Examples of robots used for interior decoration include Fireproofing

Fig. 7.15 Robot grader. A new 3-D blade control system principle for road grader (Heikkilä & Jaakkola, 2002)

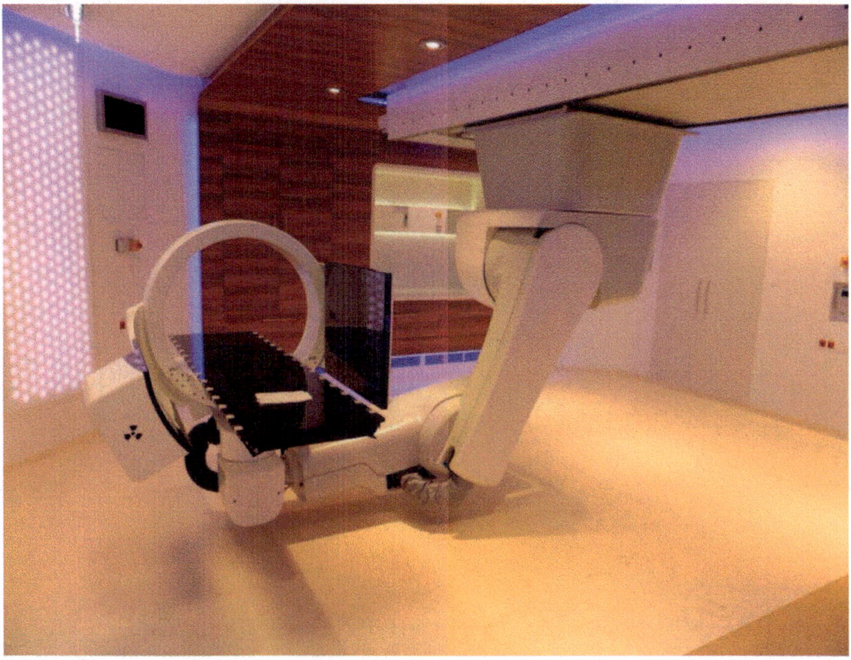

Fig. 7.16 Ceiling mounted robot (Stock et al., 2015)

spray robot (SSR-3), Ceiling-panel positioning, RoboPainter, Spray-coating robot Shimizu Ceiling Panel Robot (CFR-1), Pictobot, Quicabot, Wall-finishing robot, Steel-beam positioning manipulator, etc. (Fig. 7.17).

Fig. 7.17 PictoBot robot (automated spray painting of multiple vertical strips) (Asadi et al., 2018)

7.3 Case Studies of Robotic Construction Applications in Construction Projects and Resultant Benefits

7.3.1 Case Study One: Semi-Automated Mason (SAM or SAM100)

In the Modern era, SAM100 is boasted to be the first commercially available bricklaying robot for onsite construction activities (Dakhi & Lafhaj, 2017). Construction Robotics designed and engineered the robot and has successfully passed the prototyping phase and is now commercially available (Dakhi & Lafhaj, 2017; Madsen, 2019). SAM100 was formally launched in 2014, and since then, it has started commercial operations (Madsen, 2019). The robot has been lauded for its contribution in terms of cost and, most especially, time-saving (Fig. 7.18).

SAM 100 is not a fully automated bricklaying robot, and its performance varies based on the construction site. The robot picks up the bricks, applies for the mortar and positions the bricks on the wall but still requires a human to work alongside to smooth over the mortar before additional bricks can be placed. SAM100 costs around 500.000$ (Dakhi & Lafhaj, 2017; Madsen, 2019). According to Madsen (2019), The SAM100 has already made its way into the construction industry and has proven itself on previous projects to be an important player in improving the bricklaying process. Several companies have utilized and are utilizing this technology. Fact presented by Madsen (2019) indicated cases where SAM100 has been used, such as the Willhelm Construction in Indianapolis. The average mason lays 400 modular bricks on a good day, and after purchasing SAM100, the Willhelm Construction in Indianapolis claim they are now laying more than 2000 bricks a day. Also, Berich Masonry, a

Fig. 7.18 SAM 100 (Dakhi & Lafhaj, 2017)

Colorado-based company, began using the SAM100 at the beginning of 2018. Todd Berich, president of Berich Masonry, states that the company rents the robot at around $20,000 a month. However, Berich claims the rewards of using SAM100 are twofold, stating that the company can do more volume, or in other words, say 'yes' to clients more often (Madsen, 2019).

7.3.2 *Case Study Two: Jay and Susie Gogue Performing Arts Center Rendering*

Another case study where SAM100 was used is the Jay and Susie Gogue Performing Arts Center Rendering. During the design phase of the project, there was no intention of using the SAM100. The idea to integrate SAM100 into the project came from the masonry subcontractor's decision, and the subcontractor incurred costs relating to the robots. Having SAM100 on board, the size of the masonry crew was significantly reduced from about 15–20 masons to about 4–5 masons. Two labourers mixed the mortar, two cuts and fed the bricks into the machine, and one labourer operated the machine (Madsen, 2019) (Fig. 7.19).

The most significant benefit of using the SAM100 on the project was regarding the quality of the project as robots do not get tired; the quality of the job does not reduce. Naturally, mason's get tired at the end of the day, and the work's quality goes down due to fatigue. Using the SAM100 for the most tedious task of laying the brick increases the quality of the masonry work. In addition to quality, the SAM100 also reduces labour costs. Twelve masons would have been required to do the job in two

Fig. 7.19 Jay and Susie Gogue performing arts center rendering (Madsen, 2019)

weeks without SAM100 is done with 4 SAM100 in one week to complete the project (Madsen, 2019).

7.4 Benefits of Robotic Construction Application

Technology means more productivity with less human effort. Manual Labour has been the backbone of the construction industry, and with the application of robotics, more significant benefits accrue to the construction industry. Some of these benefits are discussed below:

(1) **Increase Efficiency**: Traditional construction industry demands a high number of manpower who are susceptible to several personal and environmental factors. However, one of the fundamental benefits of robotics technology is efficiency. Robots are designed to operate continuously without getting tired, burn out or constrained. Robots perform monotonous tasks at a faster speed and higher level of efficiency than humans. The efficiency of Robotics reflects in its impact on the speedy accomplishment of building projects, conservation of energy, environmental pollution and time consumption. More energy is expended when humans have to work for longer hours utilizing equipment powered by fossil fuel. Consequently, adoption of robotic in construction reduce emission, hence protecting and preserving the environment

(2) **Increase Quality of Construction Product**: Another advantage associated with robotic construction is that robots can perform repetitive operations, resulting in a high level of accuracy and hence impacting the quality of project deliverables. Robots can be designed to dependably and precisely repeat a task that could be monotonous and difficult for a human worker. Robots for tedious construction tasks will improve precision as robots do not experience fatigue, and human error is minimized. Also, the increase in the adoption of robotic systems will result in more automated construction parts and materials with higher quality. Consequently, the adoption of robots can leverage efficiency and speed to achieve better quality projects.

(3) **Lower Cost of Operation**: The initial cost of acquiring robots is usually high, but the cost–benefit compared to human labour is lower in the long run. The employment of diverse construction workers has cost implications such as wages, salary, incentives, feeding, fringe benefits, and medical insurance, which are not required in the use of robots. Deploying robots for construction work reduces the number of labour needed for a project, thereby cutting down costs. Most construction companies are now adopting robots to take advantage of labour and cost reduction.

(4) **Improve Workers' Safety**: Construction involves many risky tasks that expose workers to hazards resulting in injury, incapacitation, or death. Several incidences can be caused by falling, working with heavy machines, electrocution and being struck by an object. The introduction of robots to the construction

process plays a major role in promoting safety on the construction site. A robot is a machine designed like a human being that is capable of moving independently and performing complex tasks. As a result, robots can support workers in risky activities such as bricklaying, lifting heavy objects, installing heavy or large building components, taking down a wall, welding and putting down safety cones on road construction.

(5) **Reduction in material waste**: Construction generates a lot of waste during project execution. These wastes are produced from unused building materials and products of demolition such as concrete, tiles, asbestos, ceramic, wood etc. In order to minimize waste, these products can either be recycled, reused, or ensure accurate estimation of building material. However, one major characteristic of robots is creating more efficiency with limited resources than humans. Robots are designed to operate and conduct the task with minimal errors and use materials very precisely. As a result, less waste is produced. Also, they have the capability of recycling, reusing and reducing waste.

(6) **Reduce Construction Time**: Construction needs a lot of time which may run into weeks, months and even years. Manual labours are one of the leading causes of extended timing required in construction; with the adoption of robotics, the time usually spent has been largely reduced without affecting the quality of construction projects. Jobs that take two weeks with more manual labour can now be done in one week or less with less labour. Although, it is said that the set-up time of construction robots to fit into the construction environment is a significant constraint.

7.5 Risks Associated with Robotic Construction

Robotics offers several benefits to the construction industry. However, it comes with its risks and challenges. These risks are discussed below.

(1) **Environmental Risk**: Most of the challenges of applying robotic in comparison with conventional construction are the different characteristics of the environment. The implementation of robotic construction may be interrupted by harsh climatic conditions due to: adjustments concerning the environment, exposure to dust, the complexity of the environment, adjustment to changing surface conditions, interactivity between sensors and end-tools, changes like the robotized work process versus the traditional approach, human- performed work process and identification of various types of objects in natural environmental conditions (Akshatha et al., 2017).

(2) **Incompatibility of the technologies**: Incompatibility of the robots with existing practices and contemporary construction operations. It is known that humans are resistant to change, and often, workers prefer methods that they are conversant with and established solutions instead of innovative ways and technologies. This is a result of the unstable and unpredictable nature of

the construction environment. The unstructured nature of construction sites prevents the smooth integration of robotic technologies (Guglielmo et al., 2018).

(3) **Cybersecurity**: Cybersecurity is a significant threat of new technologies which are data-based in operation. This internet-connected system can be described as a means of safeguarding computer or electronic data from theft, criminals and unauthorized access. Robotics are computer-based software with components such as sensors, manipulators and control systems for effective functioning. These components act as individual computer programs that communicate with each other. The Robot Operating System integrates several hosts connected by a peer to peer topology in which messages are transmitted unencrypted. The unencrypted nature of the messages makes them vulnerable to be intercepted by an unauthorized entity (Priyadarshini, 2017).

(4) **Loss of Job**: In recent times, the robotic insurgency is transforming the world workforce, and one of the greatest fear of workers with these inventions is the loss of jobs. There has been several predictions and survey that reveals that some construction jobs will be automated. The construction industry is a highly labour-intensive sector that carries out several tedious, complex and risky tasks. With the introduction of robotics, some workers will no longer be required. Also, contractors are faced with increasing demand for better quality from clients, tighter project schedules, dwindling skilled personnel and, as a result, are opting for new technology such as robotic construction to achieve to improve their productivity.

(5) **Health and Safety Risk**: It has been established that robotics promote the health and safety of workers as workers can work in hazardous environments, and risky tasks can be conducted with robots' help. Nevertheless, robotics comes with its risks. This risk can arise from an interaction between the robots and humans as the operations of some robots demand human input. For example, miscommunication, inadequate understanding of the control of the robots can lead to injuries and even the death of a worker. Also, safety risks from robotics can be caused by electromagnetic interference (EMI). This is a situation of interference between electromagnetic fields resulting to malfunctioning of the robotics such as unexpected movement of the robot, recurrent system restarts and false triggering of a safety mechanism. To avoid these occurrences, robot developers need to understand the EMI and specify the precautionary measures.

7.6 Relevant Skills for Robotic Construction

The era of industrialization and digitization offers a wide range of technological development. These technological developments must be matched with necessary skills for effective implementation of the technologies. Research carried out by Bademosi et al. (2019) showed that the required skills needed in robotic construction could be categorised into computing and computer science. These include the following.

7.6.1 Computing Skills

These are basic computing skill sets that a modern construction engineer should get. Computing skills include;

(i) **Building information modelling**: Building information modelling (BIM) seems relatively straightforward, but it does have some common misconceptions. BIM is a design tool and software that integrates that 3D modelling with data collection. Combining the two processes provides the measure or standard for the building project—an analogy or smaller-scale representation of the final appearance and effect.

(ii) **Cloud computing**: Cloud computing refers to the storage and processing of data and software storage on remote servers rather than on-site hardware. The remote servers are housed on massive server farms run by firms such as Google and Amazon, which rent out space to store data and run applications. Computer users, in general, are beginning to transition from on-site to cloud computing, spurred in part by the lower cost of storing massive amounts of data ("big data") remotely compared to on-site. Cloud computing is expected to become an integral part of the supply chain and in-house manufacturing processes, facilitating manufacturers' ability to collect, store, and analyze large amounts of data about their production processes.

(iii) **Collaborative Environments**: Collaborative work environment or teamwork is a critical skill responsible for integrating more than one function to carry out a greater goal. In construction, irrespective of the unit, it is expedient to collaborate with other units, individuals or machines for optimal output. In robotic construction, a robotics pilot or engineer must understand the need for collaboration amongst different components of construction work. A robot must be driven or automated to fit into an environment where it'll not only function as a unit but as a member of a team working together with other robots or humans. Collaborative skills lead to knowledge and experience Integration of more than one individual, leading to synergetic creativity, innovation, problem-solving, and productivity.

(iv) **Electronic Communications**: There is no effective teamwork without proper communication. In robotic construction, various communication media are necessary: a robot must communicate with its environment using electronic media. Electronic communication is the meaningful exchange of information at a distance through technological means, mainly through electrical signals or electromagnetic waves. An in-depth or advanced electronic communication skill is required to match the highly technically driven construction industry. Electronic communication between robots and their work environment, between robots control station (i.e. for robots remotely controlled and base station for monitoring), must be professionally taken care of to avoid navigation and information glitch.

(v) **Programming**: All works need basic rules and principles for functionality. Likewise, all computers (including construction robots) also need basic rules and regulations for functionality. Giving computers rules or a set of instructions isn't an easy task, and it involves a process called Programming. Skilled in programming means a person has learned the language computers understand and can give instructions to machines to follow. Programming is creating a set of instructions that tell a computer how to perform a task. Construction robots must: move (aerial or on land), perform tasks (bricklaying, spraying, coating, welding etc.), and give feedback when necessary: these cannot be possible without a computed set of instructions that can only be done through programming, for a construction robot to work, programming skills must be present to make it happens. Programming skills are needed to create specialized engineering software to match robotic technical requirements.

7.6.2 Computer Science Skills

These are advanced computing skillsets that a modern construction engineer should get. Computer science skills include:

(i) **Algorithms**: An algorithm is simply a set of steps used to complete a specific task. They're the building blocks for programming, and they allow the likes of computers and smart devices to function and make decisions. Algorithm skills create basic programming blueprints for computerized robotic action.

(ii) **Big data analytics**: Big data analytics uses advanced analytic techniques against huge, diverse data sets that include structured, semi-structured and unstructured data from different sources and in various sizes from terabytes to zettabytes. Big data analytics examines large amounts of data to uncover hidden patterns, correlations and other insights.

(iii) **Computer graphics**: Computer graphics is routed mainly in image processing and visualization, such as two dimensional (2D), three dimensional (3D), and animated graphics. Computer graphics is the use of computers to create and manipulate pictures on a display device.

(iv) **Database concepts** (computer-based databases): Database is a storage device that keeps information for later use. Typically, a database may be a clouds base or a local base. A cloud-based database uses a non-visible local physical depository (saves data online through internet devices); a local-based database uses local physical hardware such as a hard drive, flash drive etc. Both cloud base and local database use storage devices, with the only significant difference being that cloud base use Internet services to connect to remote storage. In contrast, the local base has used resident native storage like the physical hardware of Hard disk to store data and information. A database is an organized collection of data, generally stored and accessed electronically from a computer system. A database must be integrated for interaction application and client use. This integration and interaction can be achieved through the database management system (DBMS). The database management system (DBMS) is the software responsible for the interaction between the end-users, applications, and the database itself to capture and analyse the data. The DBMS software additionally comprises all core facilities made available to aid proper use of the database.

(v) **Data structures**: The data structure is a data organization, management, and storage format that enables efficient access and modification. More precisely, a data structure collects data values, the relationships among them, and the functions or operations that can be applied to the data. Data structures keep the database organized, easy to interpret, easy to modify and developed.

(vi) **Machine Learning**: Machine learning involves computers or machines discovering how to perform tasks without being explicitly programmed to do so. It involves computers learning from data provided so that they carry out specific tasks. Machine learning is a computer algorithm related to Artificial Intelligence that helps computerized machines such as robots carry out tasks based on data obtained from sites that weren't programmed. Machine learning gives a computerized machine the ability to read, learn and carry out actions based on new knowledge.

7.6.3 *Soft Skills*

(i) **Critical thinking**: The construction site is unique and different based on environmental factors like soil, land mass, access to the construction site, the shape of a construction site, and the type of design to be erected on the construction site. An expert in robotics for construction must be a very sharp thinker—capable of carrying out logical and informed evaluation and objective analysis of construction situations to know how best fit a robot operation could fit.

(ii) **Problem Solving**: Due to the dynamic state of the construction industry, problem-solving skills for the successful adoption of robotics, flexible solutions to work site complexities must be provided to fit specific needs. Providing effective solutions to construction complexity isn't an easy task and requires talent and professionalism. A designated construction robot might have to be redesigned to suit the on-ground construction situation: for that to be possible, a robotic engineer or user must be professionally vast, experienced and possesses high intellectual capacity all running together to provide a solution. Good problem-solving skills would involve: identifying a problem and its source, device several alternative counter measures, choosing the best counter measure based on critical analysis and evaluation, and using creativity and intelligence to adopt or adapt a well-designed solution that best solves an identified problem.

(iii) **Teamwork**: The construction industry is filled with different fields of expertise that must collaborate for success. Teamwork is a collaborative effort of a unit or group to achieve a specified goal or to complete a task most effectively and efficiently successfully. Good teamwork skills must include commitment, accountability, communication and cohesion. A robot expert must know that working together with other units as a team to paramount to achieving the final construction goal. Good inter and intra unit communication and interaction skulls must be possessed by robotic expertise. The willingness to collaborate with others and accept their ideas can spring up fresh concepts or solutions.

7.7 Summary

Robotics have contributed to the improvement of the construction process in all aspects. Over the years, the construction industry has been with an image of hard work of long hours in a hazardous working environment. However, the introduction of automation and robotics technology has brought about a solution to this problem. Robotic systems have become common in many manufacturing and production operations because they have proven to be more robust, safe, efficient, accurate and productive. In the construction industry, robot technology is increasingly being adopted to build construction sites, efficiently reduce construction time, improve working conditions, and increase safety by replacing humans in dangerous operations. Automation and robotics in construction became necessary to resolve some of the problems associated with the construction industry, such as lack of skilled personnel, decreasing quality and productivity, occupational safety and substandard working condition. So far, the application of robotics initially promised relative benefits on books, but actually, the application has proven its success beyond doubt. Of all construction works, it can be said that monitoring and bricklaying sections are the two most rapidly growing sections of the robotic application with the use of Drones or UAVs and Mason Robots (SAM100 and Hadrian). Other sections like painting, welding, interior decoration, onsite transportation of heavy materials etc., have also seen robotic development capable of real-time work transformation. Some benefits of using robotics in construction include increased efficiency, increased construction quality with quality sustainability from start to finish with no varying standard, lower cost of operation, improved workers' safety and reduction in work hazards, waste, and time-saving others. Adopting robotics also has risks ranging from loss of jobs in the construction industry, cybersecurity, incompatibility of robotic technology with the different construction sites. The advent of new technologies is changing both onsite and off-site construction practice and increasing the need to acquire more engineering and cyber-related expertise and training. To fully adopt the technology, an existing practitioner must reskill and upskill to match the knowledge required to operate optimally. Cyber robotic expertise must be inculcated into the institutions training curriculum to ensure future development. Skills required may include problem-solving, critical thinking and teamwork, computing, programming, data analytics, cloud computing, electronic communication and machine learning.

References

Ahsan, A., Li, B., & Chen, I. (2018). Pictobot: A cooperative painting robot for interior finishing of Industrial Developments. *Journal of IEEE, Robotic and Automation, 25*, 82–94.

Akshatha, D., Vimala, M., Sahana, S., & Manjula, M. (2017). Robotics in construction technology. *International Journal of Advance Research in Science and Engineering, 6*(10), 547–553.

Anudari N. (2019). *Robotics in construction of concrete and masonry structures.* A thesis submitted in partial fulfilment for the degree of Bachelor in Civil Engineering branch of Building Structures 2019 Czech Technical University in Prague Faculty of Civil Engineering.

Asadi, E., Li, B., & Chen, I. M. (2018). Pictobot: A cooperative painting robot for interior finishing of industrial developments with high walls, *Article in IEEE Robotics & Automation Magazine.*

Bademosi, F. M., Tayeh, R., & Issa, R. R. (2019). Skills assessment for robotics in construction education. In *CEUR Workshop Proceedings*, 1–10. Retrieved from http://ceur-ws.org/Vol-2394/paper37.pdf.

Bock, T., Linner, T., & Ikeda, W. (2012). Exoskeleton and humanoid robotic technology in construction and built environment, The Future of Humanoid Robots. *Research and Applications, InTech.*

Bolivar, N. (2019). *Construction robotics.* Arcler Press.

Cieslaki, M., Dzialak, P., & Odyjas, P. (2015). Innovative methodology of testing of electrical remotely controlled demolition machines, Book Chapter.

Dakhi, Z., & Lafhaj, Z. (2017). Robotic mechanical design for brick-laying automation. *Cogent Engineering, 4*, 1–22. https://doi.org/10.1080/23311916.2017.1361600

Delgado, M. D., Oyedele, L., Ajayi, A., Akanbi, L., Akinade, O., Bilal, M., & Owolabi, H. (2019). Robotics and automated system in construction: Understanding industry-specific challenges for adoption. *Journal of Building Engineering, 26.* https://doi.org/10.1016/j.jobe.2019.100868.

Guglielmo, C., Alfredo, A., Alessandro, B., Marta, N., & Matteo, R. (2018). Robotics in the construction industry: state of the art and future opportunities. In *35th International Symposium on Automation and Robotics in Construction.*

Heikkilä, R., & Jaakkola, M. (2002). The efficiency of a 3-D blade control system in the construction of structure layers by road grader-automated design-build of road construction in Finland. In *Proceedings of the 19th ISARC 2002, 23–23 Sept. 2002.*

Jayaraj, A., & Divakar, H. (2018). Robotics in construction industry. In *IOP Conference. Series: Materials Science and Engineering.* IOP Publishing. https://doi.org/10.1088/1757-899X/376/1/012114.

Keerthanaa, P., Jeevitha, K., Navina, V., Indira, G., Jayamani, S. (2013). Automatic wall painting robot. *International Journal of Innovative Research in Science, Engineering and Technology, 2*(7).

Krom (1997). Industrialization and robotics in construction. *Journal of Construction Engineering and Management, ASCE, 111*(3), 260–280.

Li, R., & Ng, P. (2017). Wearable robotics and construction workers' safety and health. In *8th International Conference on Applied Human Factors and Ergonomics.*

Li, R. Y. (2018). Robots for the construction industry. *Research Gate.* https://doi.org/10.1007/978-981-10-5771-7_2

Madsen A. J. (2019). The SAM100: Analyzing labour productivity. California Polytechnic State University-San Luis, Obispo https://digitalcommons.calpoly.edu/cgi/viewcontent.cgi?article=1275&context=cmsp.

Mahbub, R. (2012). Readiness of a developing nation in implementing automation and robotics technologies in construction: A case study of Malaysia. *Civil Engineering and Architecture, 6*(7).

Malakhov, S., Bezyazeekov, P., Fedorov, Y., Kazarina, Y., Kostunin, D., & Lenok, V. (2020). Background identification algorithm for future self-triggered air-shower radio arrays. *Journal of Instrumentation, 14*,(09). https://doi.org/10.1088/1748-0221/15/09/c09026.

Oesterreich, T. D., & Teuteberg, F. (2016). Understanding the implications of digitisation and automation in the context of Industry 4.0: A triangulation approach and elements of a research agenda for the construction industry. *Computers in Industry, 83*, 121–139.

Priyadarshini, I. (2017). Cyber security risks in robotics. In *Detecting and mitigating robotic cyber security risks* (pp. 333–348).

Ruggiero, A., Salvo, S., & Laurent, C. (2016). *Robotics in construction.* Unpublished thesis, Worcester Polytechnic Institute.

Semra, A. S., Hafize, B. (2017). Robotics in architecture. In *Digital Proceeding of ICOCEE – CAPPADOCIA 2017 New Construction Methods and Hybrid Tectonics*, May 8–10, 2017.

Son, H., Kim, C., Kim, H., Han, S. H., & Kim, M. K. (2009). Trend analyses of research and development on automation and robotics technology in the construction industry. *KSCE Journal of Civil Engineering, 14*(2), 131–139.

Stein, J. J., Gotts, V., & Lahidji, B. (2002). *Construction robotics.* Retrieved from http://www.ent.ohiou.edu/~tscott/EECFG/ROBOTS.PDF.

Stock, M., Georg, P., Mayer, R., Bohlen, T., & Vatnitsky, S. (2015). Development of clinical programs for carbon ion beam therapy at MedAustron. *International Journal of Particle Therapy, 2*, 474–477.

Struková, Z., & Líška, M. (2012). Application of automation and robotics in construction work execution. *Journal of Interdisciplinary Research, 2*(2), 121–125.

Yahya, M. Y., Yin, L. H., Md. Yassin, A. B., Omar, R., Robin, R. O., Kasim, N. (2019). The challenges of the implementation of construction robotics technologies in the construction. In *MATEC Web of Conferences, 266*, 05012. https://doi.org/10.1051/matecconf/201926605012.

Chapter 8
Internet of Things (IoT)

8.1 Introduction

Technology today is an inevitable necessity for organization productivity, development of industries, economic growth, increment in a human development capacity, and livelihood improvement. Also, the advent of internet connection in technology has made it the cornerstone for every industry and a necessary fulcrum for effectiveness, efficiency and optimality. Among several industries, the construction industry is a formidable pillar in which construction projects are directly proportional to the economic development of countries. However, the construction industry has been faced with inadequacies in acquisition, management, utilization and forecasting of construction data, which has led to fluctuated productivity and inconsistent business value. Therefore, with the advent of several technologies, the Internet of Things (IoT) technology can recognise, configure, and transform data to optimize construction operations, leading to high productivity, increased business value, and cost-effectiveness.

The internet of Things (IoT) is intelligence that connects all things to the internet to exchange information and communicate through information sensing devices following agreed protocols (Chen et al., 2014). The internet of things technology is the interaction of the internet with things (objects, equipment or anything) via radio–frequency identification (RFID) tags, actuators, sensors, mobile phones, etc., for a specific objective. The IoT is a concept of connecting any device with an on–and–off switch to the internet and to each other. These include everything from cell phones, wearable devices, industrial equipment, industrial facilities and anything that can be thought of Attaran (2017). It is an information technology that enables the exchange of data and information from human–human, human–things, things–things or things–human on any network, based on standard internet protocols as shown in Figs. 8.1 and 8.2.

Moreover, Urie (2019) opined that the Internet of Things is a network of embedded sensors, meters, appliances, and devices that can send and receive data about changes to their current physical state and environment over the internet. Furthermore, the

O. Adepoju et al., *Re-skilling Human Resources for Construction 4.0*, Springer Tracts in Civil Engineering, https://doi.org/10.1007/978-3-030-85973-2_8

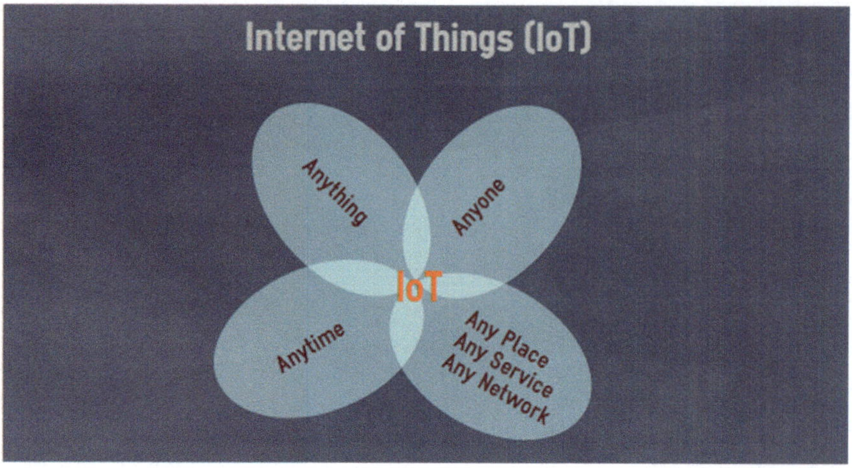

Fig. 8.1 Description of IoT (Attaran, 2017)

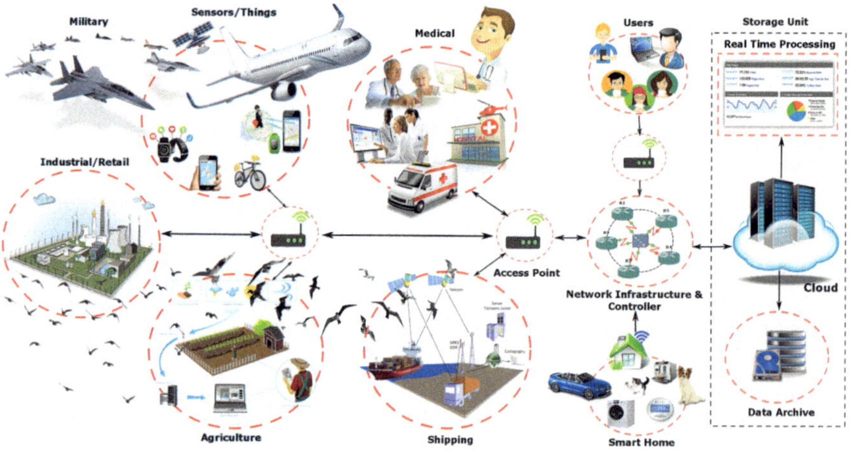

Fig. 8.2 Description of IoT and various applications (Farhan et al., 2018)

Internet Architecture Board, IAB (2015), describes the Internet of Things as "a trend where many embedded devices employ communications services offered by the internet protocols. Many of these devices often called 'smart objects' are not directly operated by humans, but exist as components in building or vehicles, or are spread out in the environment". Therefore, IoT deals with exchanging data and information between the virtual environment and the physical environment based on standard internet protocols (IP).

Internet of Things (IoT) can be enabled using various technologies, which includes Near Field Communication (NFC), Machine – to – Machine (M2M), Radio—Frequency Identification (RFID) and Vehicle-to-Vehicle (V2V) (Sindhu et al., 2016).

RFID utilizes radio waves to recognize, read and record information stored on a tag attached to a device. Tags are microchips with antennas applied to objects to obtain information using radio–frequency electromagnetic fields. NFC is wireless connectivity enabled by magnetic field induction to ensure communication between devices near each other. M2M is a process of communicating between microcontrollers, sensors, computers and smartphones, which is thus used for sensing, providing access control between different devices and processing the information. V2V technology is a wireless communication technology that enables automobiles to exchange information with one another.

The usage of IoT in the construction industry has impacted the three major phases of design, construction and management. This has led to the availability and optimization of real-time data from construction activities, which have led to: better decision making, improved health and safety of personnel, efficient performance, higher productivity, smarter designs, lesser emissions, timely project delivery and most importantly, it has led to smart construction (Gbadamosi et al., 2019). Smart construction entails utilising digital technologies throughout a typical construction project to optimise profitability, reduce construction projects life costs, sustainability, and expand clients' benefits (Reddy & Kone, 2019). The use of IoT s led to Smart buildings, which are buildings where every part of the building is run by IoT devices that automatically control the home/ building automatically, from communication devices, to entertainment devices, security, and other convenience, as shown in Fig. 8.3.

Internet of Things was first discovered in the 1980s at the Carnegie Melon University, USA, where local programmers connect through the internet to refrigerated appliances. This was done to ascertain if the drinks were available and cold before making the trip. In 1999, the term "Internet of Things" was used for the first time at a presentation for Procter & Gamble, Kevin Aston working on supply chain optimization. This was done to attract senior management attention to a new exciting technology, called Radio—frequency identification (RFID). In the presentation, the term IoT was coined to illustrate how data could be generated and processed through internet appliances without interferences of people-related problems of limited time, attention, and accuracy. He concluded that RFID was a prerequisite for IoT. This concept later gained wide acceptance and publicity in 2010 when Google's Street View stored tons of data on people's Wi-Fi networks. This generated the debate about Google's strategy to index the internet and the physical world. By 2012, Europe's most prominent internet conference, LeWeb called their theme "Internet of Things". Since then, the Internet of Things, has been gaining ground, and it's now the cornerstone of every industry, including the construction industry.

8.1.1 Components of Internet of Things

There are various components of the internet of things that works together to form a unified system. These components are categorised under sensors, data

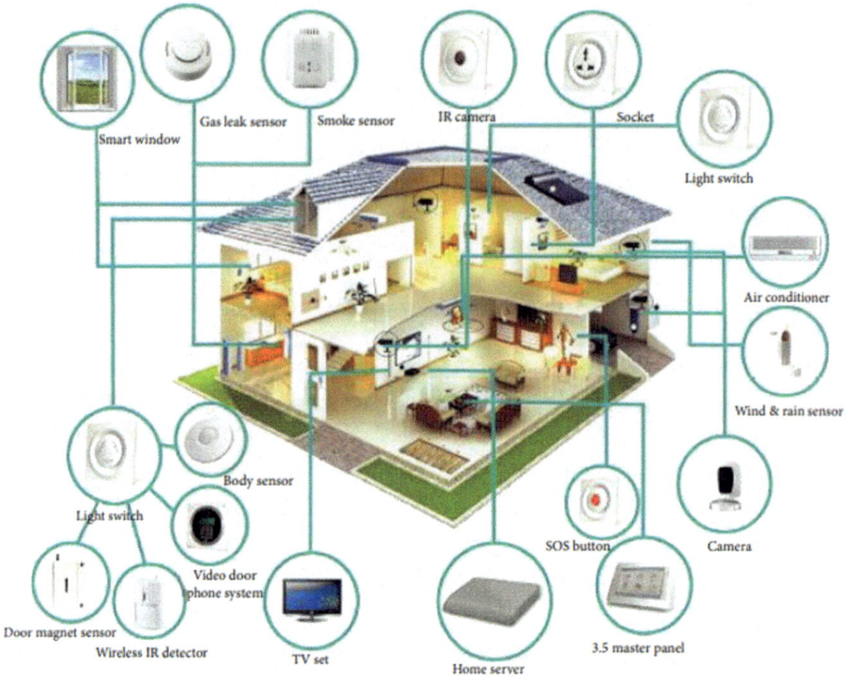

Fig. 8.3 Smart Building (Yang et al., 2018)

processing, connectivity and user interface. According to Hassan et al. (2020), the main components of IoT can be classified into the following:

(i) **Internet of Things Sensors**: Sensors known as detectors are automated devices that produce electrical, visual, or digital data generated from a physical condition or occurrence. This data produced is transformed electronically by another device into information that can be used in decision making. Sensors collect and detect relevant data from the surrounding environment. Often, they are linked to the cloud, where they can transmit and receive data. There are small/simple sensors that measure simple things like blood pressure and large sensors that can be used for advanced and complex applications.

(ii) **Actuators**: Actuators in the internet of things perform functions that are opposite to sensors. This distinction shows that actuators perform an act while sensors show the intelligence of the senses. In essence, an actuator receives a signal and perform an action accordingly. Also, these devices convert an electrical signal into a corresponding physical substance such as sound, force and movement. Actuators can be used to switch a device on or off and can regulate valves. There are several types of actuators; a common example is the servo motor which can rotate any desired degrees based on the need.

(iii) **Internet of Things Cloud**: IoT cloud represents a device designed to collect, process, store, and manage massive amounts of data. The cloud is the facility where data are processed. It can be further described as the brain of IoT that processed, command and analyse data. The cloud system is essential because IoT create enormous data from devices and applications, and as a time-sensitive device, data are required to be processed on time. With IoT cloud, information is made effectively available for clients, improvement of product and services, decisions making and preventive measure

(iv) **Internet of Things Gateway**: This is an essential component of the internet of things that connects the Internet of things devices to the cloud. It acts as a bridge that creates a way to transmit raw data from the sensors to the cloud and sent it back to the sensor's devices. It proves a cloud hub to the IoT devices and provides command, management, and control of the devices. It represents an important channel of continuous communication network between devices in the IoT. Gateways perform several functions, including pre-processing of data from the sensors before transmitting it to the cloud, translation of different network protocols to ensure interoperability of connected devices and sensors, and protecting IoT from illegal access. Importantly, gateways offer effective and easy management of data traffic in IoT.

(v) **Internet of Things Analytics and Data Management**: One of the key benefits of the Internet of things is smart analytics, which helps professionals quickly identify flaws in data collection and proffer solutions. Analytics is converting analogue data from devices and sensors into useful insights that can be understood and used for comprehensive analysis. Analytics translate raw data into valuable information that can serve as great insights in predicting future business opportunities and decision making. Smart analytics solutions are inevitable for IoT systems for management and improvement of the entire system.

(vi) **End-User Devices and User Interface**: The user interface is a component of IoT that allow a user and a computer system to interact. It constitutes the elements that a user sees and interact with; it is a device controlled by users and easily accessible. These elements include hardware, websites and applications. For example, designers use a user interface for interactive design to ease complex tasks and make them accessible to users. This device enables users to set their preferences and control of the system. A user may interact with the system through the device itself or remotely through tablets, smartphones and laptops.

8.1.2 Technologies of Internet of Things

Internet of Things exploits various networking technologies to collect, transmit, analyse, measure, and control information. These technologies make it possible to bring physical objects into the domain of cybers world. The major technologies include the following:

(i) **Internet Protocol (IP)**: Internet Protocol (IP) forms the basis for data communication over the web. It is the key network protocol used on the internet. It is a set of rules for directing and addressing the pack of data to ensure the data pass through the networks and arrive at the precise destination. There are two versions of IP which have different addresses. These include IPv4 and IPv6.The authentic protocol provides for 4.3 billion IPv4 addresses while the IPv6 will significantly supplement the availability to 85,000 trillion addresses (Madakam et al., 2015)

(ii) **Radio Frequency Identification (RFID)**: Radio Frequency Identification (RFID) technology plays a significant role in the Internet of things. It is a system that transmits the identity of an entity wireless with the use of radio waves in a serial figure format. RFDI identifies objects, collects data regarding the object, and enters the data into a computer system with little or no human intervention. One of the main benefits of RFDI is that it helps solve object identification cost-effectively. The three main components of RFDI are the tags, antenna and reader, while the reader converts the radio waves to a more usable form of data. Information collected from the tags is then transferred through a communications interface to a host computer system, where the data can be stored in a database and analyzed later. The tags are used to transmit data to the reader. Other components include software, server and access controller

(iii) **Electronic Product Code (EPC)**: Electronic Product Code (EPC)is a supporting device of RFID used to identify the world's physical objects. It is embedded in RFID to check the physical identities of objects and track them. EPC is used for sharing data in real-time by discovering a unique identifier and creating the smart industry for standard global for EPC global network. EPC is the 96–bit number linked with an RFID tag for identifying a specific tag among other tags. It differentiates two identical products and provides information such as the product's manufacture name, date and batch number.

(iv) **Barcode**: Barcode is a device that connects consumers with the internet of things through a mere scan of communication devices. In contemporary information technology and communication systems, information is often encoded. These codes are used for identification and security. Barcode is a method of encoding numbers and letters using bars and spaces of different sizes. They can be described as a printed sequence of parallel lines with varying widths. These lines are processed and entered into a computer system, and the bars are often black colour on a white background. The codes are applied to products for easy identification. A key benefit of barcode systems is that it allows users to process detailed information when the barcode is scanned other than merely storing data for future processing.

(v) **Bluetooth**: Bluetooth is a significant short-range communication device and protocol for internet of things applications. It acts as a communication layer and data link which connects sensors to the gateway. It is a wireless technology that makes it possible for wireless connectivity, and in a situation of no wireless facility, Personal Area Networks can be designed. Comparing Bluetooth to other communication protocols such as WiFi, RFID, WAN, and NFC is cost-effective.

(vi) **ZigBee**: Globally, Zigbee is an internet of things protocol that is widely accepted, like Bluetooth. It is also a wireless technology that allows devices to communicate through several communication networks. This is achieved using a mesh network, which will enable devices to communicate with several networks. According to Madakam et al. (2015), ZigBee has a range of around 100 m, and a bandwidth of 250 kbps and the topologies that it works are mesh, star and cluster. Zigbee is specifically designed to meet the needs of low cost and low power of internet network. Examples of appliances that Zigbee is applied include lighting control, sensors, security and other home technology with low power consumption. Consequently, it is considered the most standard protocol in wireless communication based on unique characteristics such as flexibility, reliability and long-lasting battery life span.

(vii) **Near Field Communication (NFC)**: Near Field Communication is a technology designed to enhance the effective functioning of the internet of things where physical objects of all kinds can collect and exchange data. NFC is essentially a short-range high-frequency Radio Frequency Identification (RFID) technology that allows exchanging information between Near Filed Communication enabled devices. These devices can communicate to various applications such as communication between smartphones and booking of a ticket at the airport and bank payment using NFC credit card for security. It is also a wireless, low data communication technology that uses electromagnetic waves.

(viii) **Wireless Sensor Networks (WSN)**: Wireless Sensor Network (WSN) is an essential technology as sensors are considered the Internet of Things' key technology. WSN is a network that comprised several sensor nodes, with each node being equipped with a sensor to detect physical occurrences. Sensors nodes is a major part of WSN, which screen data collected before transmitting to other nodes. WSN enabled interaction between computers and person, senses and also controlled the immediate environment.

8.1.3 Architecture Layers of Internet of Things

According to Attaran (2017), four (4) major layers of IoT architectures regulate IoT devices' internal workings. The layers are;

(a) **Sensor, Connectivity and Network Layer**: this is the bottom of IoT Architecture, which constitutes sensors, RFID tags and a connectivity network that collects information. The RFID tags, barcode reader, and sensors are essential parts of an IoT, and they are wireless devices responsible for collecting raw data. Simply put, they are the messengers of IoT devices. These essential parts work within network connectivity like Wide Area Network (WAN), Personal Area Network (PAN), etc., and thus communicate the raw data to the next layers.

(b) **Gateway and Network Layer**: this layer ensures that data coming from the bottom layers are routed correctly to the next layer through a set of networks and gateways. The gateways include signal processors, embedded operating systems, microcontrollers and gateway networks are Local Area Network, LAN, and Wide Area Network, WAN. This layer has a large storage capacity to store an enormous amount of data from the bottom layer.

(c) **Management Service Layer**: this layer processed the information from the gateway and network layer through the following activities; text mining, data mining, service analytics analysis of IoT devices, information analysis, and device management.

(d) **Application Layer**: the processed data are then used for various applications using different services based on suitable communication models.

8.1.4 Communication/Connectivity Models for Internet of Things

According to the Internet Architecture Board, IAB (2015), and Kulkarni and Kulkarni (2017), there are four common communication models used by IoT devices, which makes the difference between different IoT devices. They include:

(a) **Device–to–Device (D2D) Model**: This model is two or more devices that directly exchange information through different types of networks that comprise internet protocols. These connections may be gotten from protocols such as Bluetooth and Z—wave. This model, as depicted in Fig. 8.4 is usually used in devices found in buildings such as light bulbs, lights switches, thermostats, door locks, etc.

(b **Device–to–Cloud Model**: This model is designed in the form of service providers that connect directly to an internet cloud service for data exchange and relevant information, as shown in Fig. 8.5. The internet cloud services are enabled through mechanisms such as wired Ethernet or wifi connections. This mechanism establishes a connection between the device and IP networks such

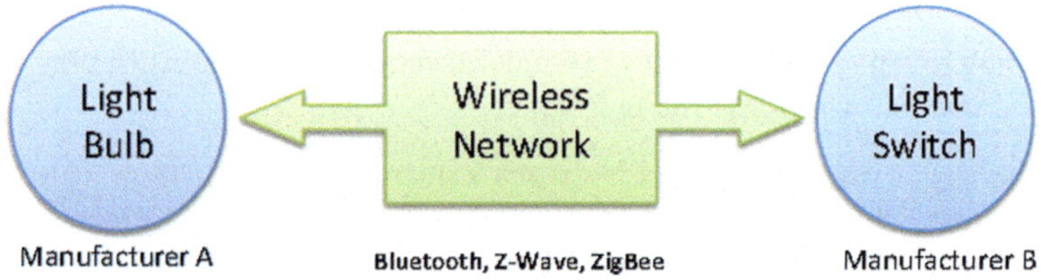

Fig. 8.4 Device to device model (Kulkarni & Kulkarni, 2017)

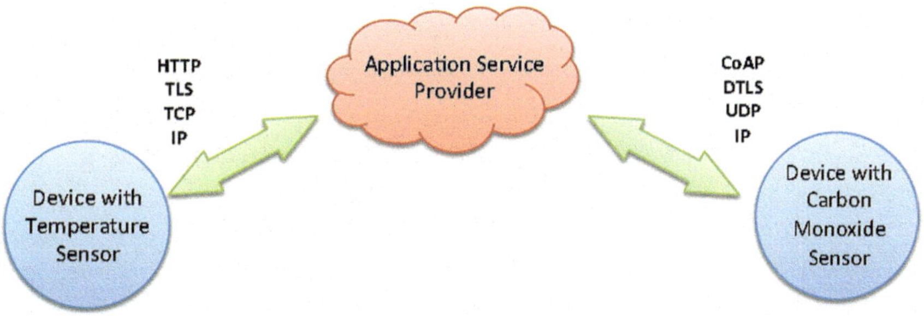

Fig. 8.5 Device to cloud model (Kulkarni & Kulkarni, 2017)

as Transmission Control Protocol (TCP), Hypertext transfer protocol (HTTP), Constrained Application Protocol (CoAP), Datagram transport layer security (DTLS), Transport layer security (TLS), and User datagram protocol (UDP). Comparing this second model to the first model, the device to cloud connectivity is not limited by distance but based on regulations and the relationship between the service provider and the user.

(c) **Device–to–Gateway Model**: This model involves using an intermediary device by an IoT device to access a cloud service. The intermediary devices are usually a software application that acts as a gateway to a service provider as shown in Fig. 8.6. The gateway in this model provides security, unique functionality such as data or protocol translation. The gateway device is a smartphone that runs an application to communicate with a device and relay such information to a cloud service (Fig. 8.6).

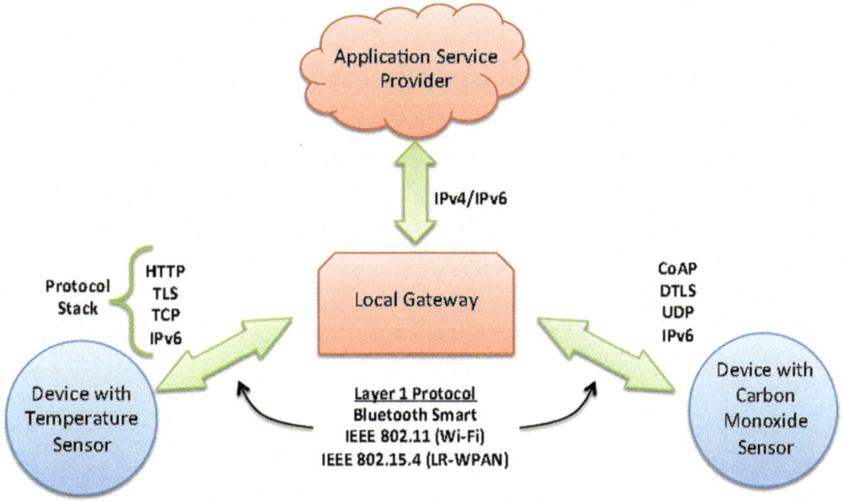

Fig. 8.6 Device to Gateway model (Kulkarni & Kulkarni, 2017)

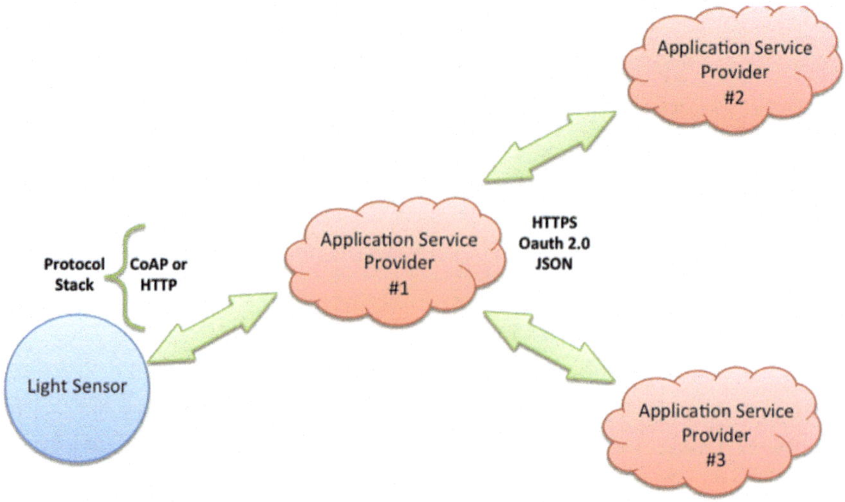

Fig. 8.7 Back-end data sharing model (Kulkarni & Kulkarni, 2017)

(d) **Back–End Data Sharing Model**: This model optimizes the single device to cloud communication model so that Authorized third parties can access ioT devices and sensor data. The Back–end data-sharing model enables users to export and analyze smart objects data from a cloud service in combination with data from other sources. As shown in Fig. 8.7, this communication model suggests a federated or central cloud service approach or cloud applications programmer interfaces (APIs), which are all needed to achieve interoperability of smart device data hosted in the cloud.

8.2 Applications of Internet of Things to the Construction Industry

The application of IoT in the construction industry is necessary for the optimization of construction activities. This ensures high construction management performance, encompassing project objectives without time or cost overruns and the high quality of projects (Mahmud et al., 2018; Reddy & Kone, 2019; Urie, 2019). The applications in the construction industry are divided into two categories; on-site applications and off-site/construction buildings applications.

8.2.1 On-Site Applications

Almost every construction project starts from the construction site with many activities. The revolving around the implementation of construction designs, which thus necessitated the involvement of technology such as the IoT device. The application of the IoT technology on-site can be in the following ways:

(a) **Digital Communication**: many devices are IoT oriented, which has aided communications among construction workers, construction project stakeholders and construction activities. The use of social media such as WhatsApp, Facebook, Telegram and email on smartphones have made the communication of project status, project progress and project monitoring effective and stress-free. The use of these applications on smartphones, tablets, and computers (all IoT enabled devices) has easily exchanged opinions, video calls, and timely information. Also, the use of Scan marker air or pen scanner, a digital pen that scans any printed texts and transmits the text into any device via Bluetooth connections. This IoT device can be used to transmit any construction documents/ texts into the computer or Smartphone, or tablet. It can also translate around forty (40) languages and text scans.

(b) **Project Monitoring**: to ensure smoothness of the project on the construction site, projects must be monitored effectively. Hence, IoT devices like drones are deployed to the site to monitor activities, project stages, security purposes, and employees well-being, as shown in Fig. 8.8.

(c) **Material control**: project smoothness is directly proportional to the control of materials on site. Using the IoT to ensure that the right quality and quantity of materials are on site: the supply unit of the materials can be labelled with a Radio frequency identification (RFID) tag, which would ensure that materials are counted automatically. They provide information when they fall below

Fig. 8.8 A drone used to monitor a construction site (Tkac & Mesaros, 2019)

Fig. 8.9 Construction materials being identified and counted at construction site gate (Patil & Shelake, 2021)

expectation, as depicted in Fig. 8.9. Also, the RFID tags ensure that materials are easily accessible, the suitability of the temperature of the material are verified, the extent of material damage and that materials are restocked when depleting.

(d) **Maintenance of Machinery and Equipment**: one of the main components of an IoT device is a sensor, which forms part of the bottom layer of IoT Architecture. This sensor can also be connected to machines and equipment in the construction industry, especially heavy construction machinery. The sensor monitors from a distance necessary maintenance required and condition monitoring, such as temperature range, vibration level, ductility level, etc., as shown in Fig. 8.10. This will ensure that there are no avoidable damages, additional costs and unnecessary delays.

(e) **Power, Energy and Fuel (PEF) savings**: through the IoT approach, information about the amount of power, energy and fuel consumed on construction sites by offices and machines can be generated, managed and saved. This is done through IoT devices, which involves automatic switching on/ off office lighting and pieces of machinery when they exceed the required period of operation. Furthermore, Motlagh et al. (2020) opined that using IoT devices, an integrated smart energy system can be generated, as shown in Fig. 8.11.

(f) **Security control**: many construction sites have stores where the materials are kept for easy access when needed to avoid delays. However, there is a possibility that these materials will be stolen if kept on the construction site. The possibility of theft, which poses a challenge, can be curbed using the IoT. A practical example of implementing IoT is when RFID tags are placed on materials. When there is a theft or attempted theft, the sensors on the materials will detect the current location of the materials or item. Also, on or off-site, an IoT device known as August Smart Lock installed on doors, as shown in

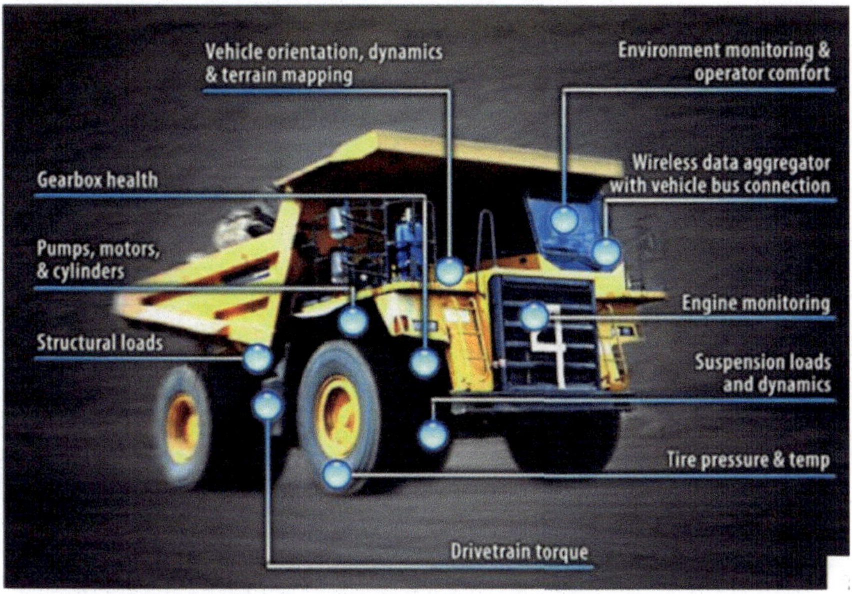

Fig. 8.10 Checking the state of heavy machinery (Jiang & He, 2020)

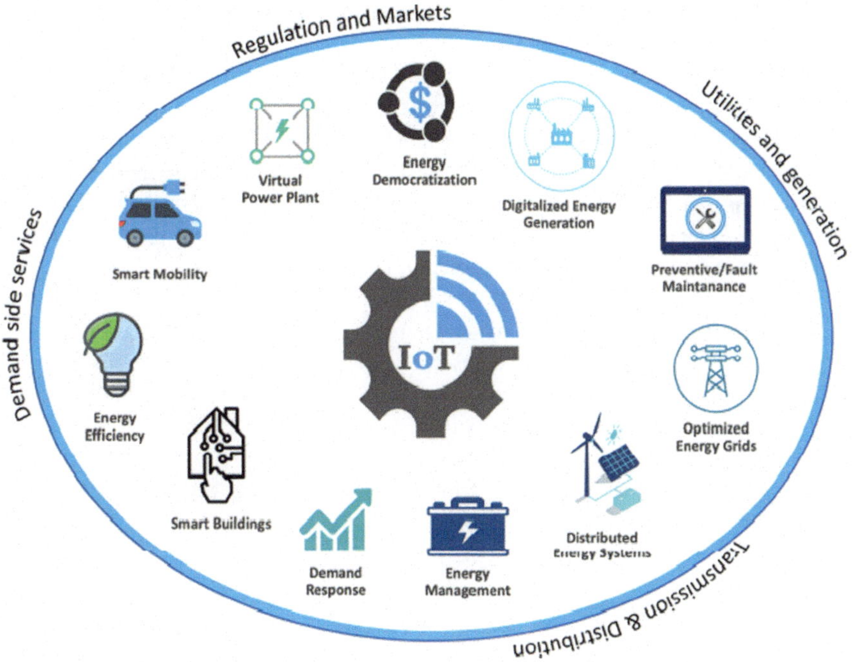

Fig. 8.11 An integrated smart energy system using IoT (Motlagh, et al., 2020)

Fig. 8.12 A smart lock (Aluri, 2020)

Fig. 8.12, can detect any image, even guest images, and trigger an alarm when a threat is detected. The device can be lock and unlock anywhere using an application on Smartphones.

(g) **Workers Management**: construction workers are one of the most vital assets on construction sites; hence, effective management is essential to the project's progress. Therefore, sensors are connected to the worker's badge through an IoT device, which contains project information about their availability on on-site, skills & qualifications, time duration on tasks, and their location on site. This information could be used effectively for payment systems, human resource management, planning, and other managerial functions. Also, workers' health can be ascertained using an IoT device that can record pulse rate, temperature rate, etc., as seen in Fig. 8.13. It can also detect when a worker is at risk. It can also be used to reduce absenteeism among workers.

(h) **Concrete curing**: in construction, sensors are embedded in concrete during casting to properly monitor and plan schedules with certainty, as shown in Fig. 8.14. The sensor provides information on the comprehensive strength of concrete, formwork removal time, opening a bridge to traffic, concrete mix design, and prestressed cable tensioning time, as shown in Fig. 8.15: all this enhanced quality and service delivery.

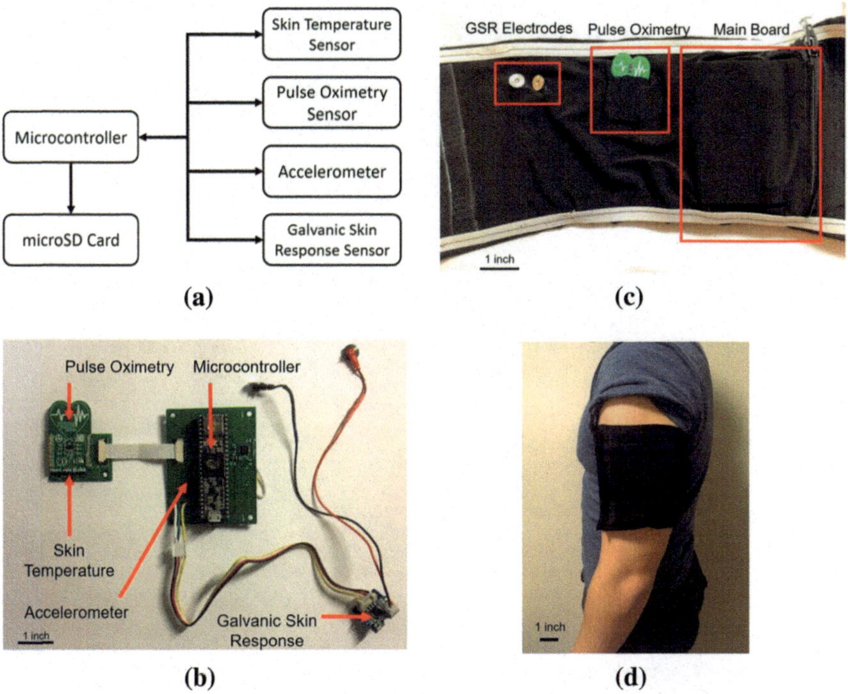

Fig. 8.13 A diagrammatic display of sensors connected to the human body (Pham, et al., 2020)

8.2.2 Off-Site Construction Buildings Applications

This application category deals with the usage of IoT IoT-enabled devices when the project work is completed and for existing buildings. They are applied in the following ways;

(a) **Environmental Monitoring**: through IoT devices such as Waspmote Plug & Sense, shown in Fig. 8.16, the environments and the buildings can be monitored. The device sensor monitors the possibility of fire outbreaks, flood threats, air quality, waste containment level, structural health of buildings, noise maps, dust concentrations, humidity, temperature, luminosity level, and electricity measurement.

(b) **Building structure Health Monitoring**: buildings have a lifecycle for maintenance and renovations periods, which exactness of period cannot be fixed. However, due to IoT devices like laser displacement sensors (which operating mechanism is shown in Fig. 8.17), structures can detect cracks, vibration, building conditions, the state of building materials, and buildings/construction structures like bridges and monuments.

(c) **Smart Buildings**: smart buildings are buildings mostly controlled by IoT enabled devices such as heating, ventilation, and air conditioning system

Fig. 8.14 A sensor connected to a cast iron (Cabezas et al., 2018)

(HVAC), energy management, lighting system, fire prevention device, crime prevention device, closed-circuit television (CCTV), sound detection, a video camera for surveillance, photovoltaic walls, smart locks, smart switches, and Aerogel sealant for air leakage/smart windows. This is depicted in Fig. 8.18. These devices in a smart building communicate necessary information to both building users and facility managers for proper regulations.

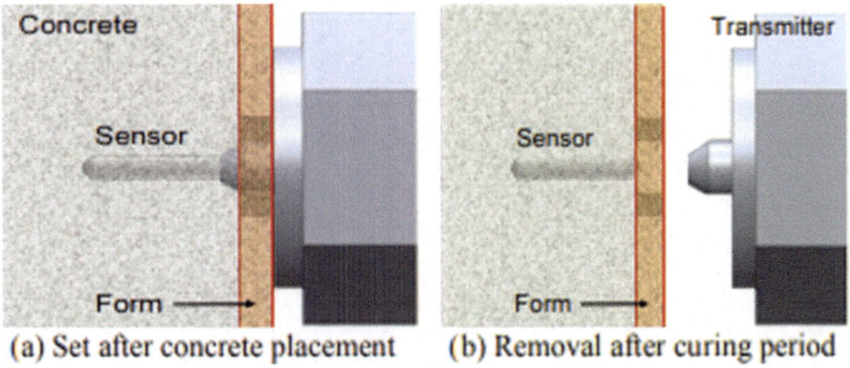

Fig. 8.15 A connected sensor for data collection in a concrete (Lee et al., 2014)

Fig. 8.16 Waspmote Plug & Sense device (Velasco et al., 2016)

8.3 Benefits of Internet of Things in the Construction Industry

The applications of IoT to the construction industry have brought needed, timely and efficient benefits to the industry, which includes the following;

(1) **Availability of Real-time data**: Through IoT devices' usage in every process and activity of a construction project, data are generated and can be managed. This available data enables informed decision-making and efficient performance monitoring.

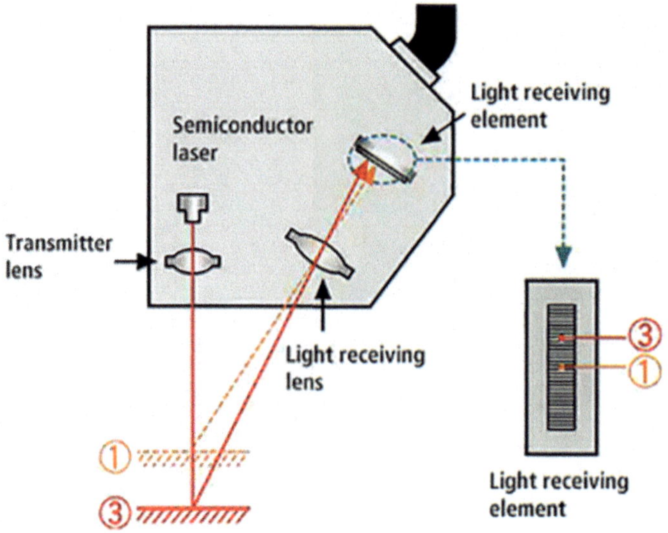

Fig. 8.17 A laser displacement sensor (Gopinath & Ramadoss, 2021)

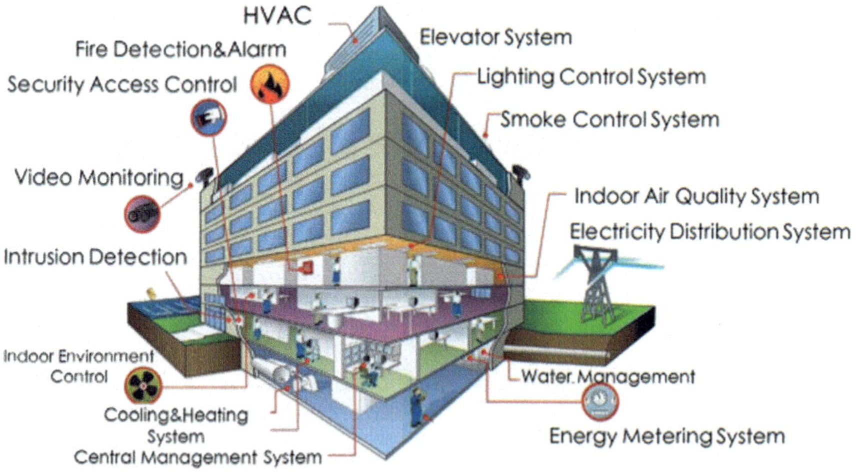

Fig. 8.18 A view of functions of IoT in a building (Zhao et al., 2018)

(2) **Ensuring preventive maintenance**: the maintenance of construction projects without IoT enables devices are usually corrective due to the absence of real-time caution. However, using IoT devices, through the use of sensors and RFID tags, vulnerabilities on construction sites or smart buildings are easily detected and prevented from causing any damages. Hence, preventive maintenance of IoT devices on construction projects optimizes project delivery.

(3) **Ensures administrative Efficiency**: in the construction industry, there are lots of exchanges of documents, memos, drawings, printings, tasks reallocation, unplanned activity lags, unforeseen lead time, etc., which rely on manual processes. However, these administrative bureaucracies are digitized with IoT devices, thus saving and optimizing processing time, sending accurate information at the right time, in the right format, and to the right recipients.

(4) **Cost-effectiveness**: throughout the world, construction projects are usually capital intensive and vary depending on the size and type of projects. However, with IoT applications, the cost is being minimized in the design stage through: tracking labour hours, waste minimizations, theft impossibility, workers' safety, maintenance of machinery & equipment, workers' effectiveness, reduction in rework, and avoidance of natural disaster mitigating effect of natural disasters. Therefore, IoT in the construction industry ensured the effectiveness of cost in the industry, even with low-cost sensors (Gbadamosi, et al., 2019).

(5) **Safety of workers**: on construction sites, many workers are performing different activities, and with IoT devices & sensors, they could be tracked and located in case of any accidents or security threats.

(6) **Security of Materials**: Construction sites are often associated with theft, wastes, and vandalism, resulting in loss of profit and insurance loss. However, construction companies are alerted about any guest image or missing material or any smoke or any accident with IoT devices.

(7) **Improvement of building lifespan**: most buildings collapse and deteriorate before reaching their lifespan due to excessive usage of energy, overloaded bodies on the building, etc. However, with IoT devices like a smart thermostat, circuit breaker, humidity sensor, linear displacement sensor, digital power meters, etc., building lifespan will increase. However, this is based on adherence to the information generated from IoT devices.

(8) **Certainty of Productivity**: IoT technologies in the construction industry will undoubtedly ensure the productivity of construction due to cost-effectiveness of the project, adequate supply of materials, adequate scheduling for delivery date, an automatic commitment of workers through effective monitoring, and the usage of quality material.

8.4 Threat Associated with IoT in the Construction Industry

The sophistication of technology makes the risks higher, and implementation challenges are unique, especially in a people-oriented industry like the construction industry. Below are major challenges that confront IoT utilizations which are applicable in the construction industry (Gazis et.al., 2015);

(a) **Technology and semantic interoperability**: The IoT is an omnipresent network of both the physical and virtual world, focusing on connecting devices. However, IoT's challenge is incompatibility or technological interoperability

of the devices that need to be networked together due to their different technological capability. This is also because many devices have their own IoT solutions that are likely incompatible with other IoT solutions. Also, IoT devices are challenged by their different semantic interoperability or information model. Generally, IoT devices are heterogeneous in terms of underlying communication protocols, data formats, different ontologies, semantic web, and technological capability.
In addressing the challenge associated with interoperability of IoT devices, the Internet Society (2015) poses the following question for consideration during the manufacturing process;

- Are the current interoperability standards sufficiently similar or different across the wide range of potential IoT applications and use cases?
- What are the generic and widely available standards (such as the internet protocol suite) that could be used as building blocks for IoT devices and services?
- How would a lack of interoperability impacts users' ability to connect, speak, share and innovate?
- What are the optimal roles of standards developing organizations (SDOs), industry consortia, and stakeholders' groups in IoT standards development?
- What is the best approach to educate and engage user and developer communities about problems of badly behaving IoT devices?

(b) **Security and Privacy Challenges**: The Internet of Things is mostly about data acquisition, processing, management, and sharing about a situation, a person, or an environment depending on the purpose of deployment. However, there is a security challenge of these data being hacked and users' privacy being breached and exchanged with unauthorized third parties. Hence, issues such as data integrity, verification procedure, unique identification, data ownership, liability and legal matters should be considered during manufacturing, installing, and transferring ownership.

(c) **Schedule Risk of Manufacturers**: technology is a high value-driven sector of any economy, which yields a high return on investment; hence, competition among companies and brands is unavoidable. Also, the advantage of first-mover in the market is a strategy in getting a sizeable market share. However, during this business strategy thinking and deployment, there is a risk of the device lacking certainty in standards development schedules and processes due to impulsiveness.

(d) **Configuration risk for users**: in the construction industry, buildings have many parts that need different IoT devices, requiring configuration settings of the different devices. A user with poor knowledge of how-to work on the configuration or a sophisticated configuration setting poses a risk for the user, the building, and the device's brand. Hence, simplicity in the configuration of IoT devices should be prioritized.

8.5 Relevant Skills for the Internet of Things

The development of IoT devices requires a pool of different abilities, knowledge, and skills (both hard/technical & soft skills). However, technical skills are mostly needed, for the operation of IoT. These skills are essential to the device, the brand, and the user. In recent times, the complexity of IoT technology is creating a high skills gap in the workforce and, hence limiting its implementation. Therefore, for organizations, including the construction industry, to succeed in implementing IoT, there is a need to get acquitted with the relevant skills. Below are major skills required for successful implementation of IoT technologies/ devices;

(a) **Business Intelligence**: This skill involves collecting, storing, and analysing data streams using a different necessary software tool. It also includes data centre management, sensor data analysis, predictive analytics, and programming in relevant languages depending on the device's purpose.

(b) **Data security**: The complexity of internet of things connectivity and communication has made security skills of great importance. Security skill is also important as Iot involves exposure to large volumes of data and connectivity with several devices, making it highly risky. This skill will help mitigate data privacy security risk and identity physical and logical threats. one of the IoT challenges is unauthorized data access, use, sharing, manipulation, interference, and destruction. Hence, data security skill is essential and encompasses vulnerability assessment ability, public key infrastructure (PKI) security, ethical hacking, and wireless network security.

(c) **Artificial Intelligence Skill**: The Internet of things generates a high volume of data. This requires a highly efficient and faster mechanism to collect, organize and analyze data. AI skills also help to filter unnecessary data

(d) **Data Analytics**: The introduction of IoT has led to the mining of data in terabytes per second, and putting this data to use is as important as the data itself. Data analytics is required to extract data and build algorithms that help to annex out insight and analytics out of data. Skill in data analytics requires knowledge of machine learning, data presentation, data visualization, programming, SQL etc.

(e) **Teamwork**: An IoT device is not a one-person special effort but a coordination of different human capital. Teamwork is highly essential for all team members to achieve the product specifications and objectives. Teamwork involves tolerance, patience, understanding of one's role and other's roles, unity of purpose, and individual/ collective coordination.

(f) **Designing**: Designing skill is very crucial in IoT especially AutoCAD. This is a premier design software experiencing an increase in demand due to growth in IoT complexity. The AutoCAD permits late changes in the product development process to be affected. Also, IoT demands application design in which devices are set in motion by humans in a situation. When there is a bad interface between the device and the user, the goal of the device will be limited. Hence, application designs skills include a user interface (UI), user experience

(UX), product designs, graphic designs, and other necessary designs vital to IoT devices.

(g) **Mobile Applications Development**: IoT devices are mostly controlled on smartphones' applications; hence, mobile applications developers are crucial to IoT devices. The developers must be sound in cloud computing, wireless technologies, mobile technologies, and different programming languages, and how the applications can work on Androids and Apple ioS.

(h) **Hardware interfacing**: IoT devices are both communication and electronic hardware, which demands the knowledge of engineers or people that are good in AutoCAD designs, wireless sensor modelling, quality management, and Microelectromechanical systems (MEMS).

(i) **Internet Protocol (IP) Networking**: routing data and information from an IoT device to another is the basics of IoT technology, which is made possible through various internet protocols networks, which must be secure, reliable, set up for enormous traffic, and perfect. The IP networking skill includes the ability to set up and resolve: modern networks, use of RFID, conversant in Wifi protected access (WPA)/ wired equivalent privacy (WEP) or 3G/ 4G networks, proficiency in existing networking practices and technologies, knowledge of Bluetooth low energy, knowledge about the basics of open system interconnection (OSI) stack, how connectivity protocols work and latest standards in IoT communication.

(j) **Machine Learning**: Machine learning helps create a smarter application and make predictions through sensors and other connected devices.

8.6 Summary

Internet of Things is the connection of anything to anything/ anyone at any time, at any place, at any service, and any network based on standard internet protocols. Its basic foundation is RFID and sensors with relevant networks. IoT, first called by Kevin Aston in 1999, has four (4) communication models; a device-to-device model, a device-to-cloud model, device to gateway model, and a back-end data-sharing model. It also has four Architectural layers; sensors, connectivity and network layer; gateway and network layer, management layer, and application layer.

Internet of Things is a landmark phenomenon for the construction industry, making: construction work efficient, easier life for construction end product users, high productivity and profitability, effective and timely maintenance of machinery and an effective decision-making process. Also, introducing the IoT in the construction industry has led to several benefits: availability of real-time data, preventive maintenance, administrative efficiency, cost-effectiveness, the safety of workers, security of materials & buildings, improvement of buildings lifespan, and certainty of productivity. However, to sustain the gain of IoT in the construction industry, there must be: effective technological and semantic interoperability of IoT devices, productive teamwork, usage of highly skilled and knowledgeable personnel in the manufacturing process, high level of data security, and patient adherence to necessary IoT standards before market penetration.

References

Aluri, D. C. (2020). Smart lock systems: An overview. *International Journal of Computer Applications, 177*(37), 40–43.

Attaran, M. (2017). The internet of things: Limitless opportunities for business and society. *Journal of Strategic Innovation and Sustainability, 12*(1), 10–29.

Cabezas, J., Sanchez-Rodriguez, T., Gomez-Galan, J. A., Cifuentes, H., & Carvajal, R. G., (2018). Compact embedded wireless sensor – based monitoring of concrete curing. *Sensors, 18*(876), 1–17. https://doi.org/10.3390/s18030876.

Chen, S., Xu, H., Liu, D., Hu, B., & Wang, H. (2014). A vision of IoT: Applications, challenges, and opportunities with China perspective. *IEEE Internet of Things Journal, 1*(4).

Farhan, L., Kharel, R., Kaiwartya, O., Quiroz-Castellanos, M., Alissa, A., & Abdulsalam, A. (2018). A concise review on internet of things (IoT) -problems, challenges and opportunities.In: *2018 11th International Symposium on Communication Systems, Networks & Digital Signal Processing (CSNDSP)*, 2018 (pp. 1–6). https://doi.org/10.1109/CSNDSP.2018.8471762.

Gazis, V., Goertz, M., Huber, M., Leonardi, A., Mathioudakis, K., Wiesmaier, A., Florian Zeige, F. (2015). Short paper: IoT: Challenges, projects, architecture. In: *Conference paper, AGT Group (R&D) GmbH, Germany*. https://doi.org/10.1109/ICIN.2015.7073822.

Gbadamosi, A., Oyedele, L., Mahamadu, A., Kusimo, H., & Olawale, O. (2019). The role of internet of things in delivery smart construction. In: *CIB World Building Congress*, 2019, Hong Kong SAR, 17–21 June, 2019.

Gopinath, V. K., & Ramadoss, R. (2021). Review of structural health monitoring for restoration of heritage buildings. *Material Today: Proceedings, 43*(2021), 1534–1538.

Hassan, R ., Qamar, F. , Hasan, M., Aman., A & Ahmed, A. (2020). Internet of things and its applications: A comprehensive survey In: *Symmetry* 2020 (Vol.12, p.1674). https://doi.org/10.3390/sym12101674, www.mdpi.com/journal/symmetry/MDPI.

Internet Architecture Board, IAB (2015). RFC 7452, Architectural Considerations in Smart Object Networking. Retrieved March, 2015, from https://tools.ietf.org/html/rfc7452.

Internet Society (2015). The internet of things: An overview. Understanding the issues & challenges of a more connected world. www.internetsociety.org.

Jiang, Y., & He, X. (2020). Overview of applications of the sensor technologies for construction machinery. *IEEE Access, 8*(2020), 110324–110335.

Kulkarni, S., & Kulkarni, S. (2017). Communication models in internet of things: A survey. *International Journal of Science Technology & Engineering, 3*(11), 87–91.

Lee, H., Cho, M., Yang, H., Lee, S., & Park, W. (2014). Curing management of early – age concrete at construction site using integrated wireless sensors. *Journal of Advanced Concrete Technology, 12*, 91–100. https://doi.org/10.3151/jact.12.91.

Madakam, S., Ramaswamy, R., & Tripathi, S. (2015). Internet of things (IoT): A literature review. *Journal of Computer and Communications, 3*, 164–173. Retrieved from http://dx.doi.org/https://doi.org/10.4236/jcc.2015.35021.

Mahmud, S. H., Assan, L., & Islam, R. (2018). Potentials of internet of things (IoT) in Malaysian construction industry. *Annals of Emerging Technologies in Computing (AETiC), 2*(1), 44–52, Retrieved from http://aetic.theiaer.org/archive//v2n4/p4.pdf.

Motlagh, N. H., Mohammadrezaei, M., Hunt, J., & Zakeri, B. (2020). Internet of things (IoT) and the energy sector. *Energies, 13*, 494. https://doi.org/10.3390/en13020494.

Patil, P., & Shelake, A. (2021). Application of RFID technology to improve material management on construction sites. *International Research Journal of Engineering and Technology (IRJET), 8*(2), 1000–1003.

Pham, S., Yeap, D., Escalera, G., Basu, R., Xiangmei, W., Kenyon, N. J., Hert-Picciotto, I., Michelle, J., & Davis, C. (2020). Wearable sensor system to monitor physical activity and the physiological effects of heat exposure. *Sensors., 20*(855), 1–15. https://doi.org/10.3390/s20030855.

Reddy, H. G., & Kone, V. (2019). Study on implementing smart construction with various applications using internet of things techniques. *International Journal of Recent Technology and Engineering (IJRTE), 7*(6C2).

Sindhu, K. S., Aasha, M., & Sivakumari, S. (2016). A study on technologies user for implementation of internet of things. *World Scientific News, 41*, 114–119.

Tkac, M., & Mesaros, P. (2019). Utilizing drone technology in the civil engineering. *Journal of Civil Engineering, 14*(1), 27–37. https://doi.org/10.1515/sspjce-2019-0003.

Urie, M. (2019). The internet of things in construction. In: *Gardiner & Theobald Publication.*

Velasco, A., Ferrero, R., Gandino, F., Montrucchio, B., & Rebaudengo, M. (2016). A mobile and low – cost system for environmental monitoring: A case study. *Sensors, 16*(710), 1–17. https://doi.org/10.3390/s16050710.

Yang, H., Lee, W., & Lee, H. (2018). IoT smart home adoption: The importance of proper level automation. *Journal of Sensors,* 1–11, 6464036, https://doi.org/10.1155/2018/6464036.

Zhao, Q., Xia, L., & Jiang, Z. (2018). Project Report: New Generation intelligent building platform techniques. *Energy Informatics, 1*(2), 1–5. https://doi.org/10.1007/s42162-018-0011-9.

Part III
Reskilling and Roles of Stakeholders for Implementation of Construction 4.0

Chapter 9
Reskilling for Construction 4.0

9.1 Introduction

The world is constantly changing on the wheels of technological advancements which demand human capital development and acquisition of relevant skills. Over the years, the world has witnessed three different revolutions and is currently in the fourth industrial revolution (4IR). The fourth industrial revolution has altered and is still altering how we live, learn, work, eat and communicate. In the world of works, this revolution greatly affects how products are imagined, manufactured, distributed, acquired, and consumed. The expectations from businesses, employees and society as a whole have taken a new turn. Ultimately, these changes are redefining job requirements. As a result, it becomes necessary to get familiar with the demands that come with the revolution for quick adaptability. The degree of adaptability depends to a large extent on the emphasis on human capital development. This demands the involvement and collaborations of the industries, government, and educational institutions towards this development. The collaboration of the stakeholders is crucial as human capital development is a real-time investment. It is thus believed that this will facilitate and motivate workers to learn and be well equipped to appropriate the new technologies. Consequently, this goal can be achieved through the implementation of upskilling and reskilling of human resources. In line with this preceding, the emergence of the fourth industrial technologies has necessitated the need to reskilling human resources to remain relevant in an ever-developing world.

The construction industry employs several categories of workers for the successful completion of a project. These include the professionals (architect, engineers, surveyors etc.), skilled (plumbers, electricians, carpenters, bricklayers etc.) and semi-skilled workers (messenger, janitor, truck driver). In retrospect, the construction industry has constantly been challenged with a shortage of skilled manpower, and the introduction of construction 4.0 is likely to widen the skill gap (Adepoju & Aigbavboa, 2020). Also, there have been several predictions that adopting these new technologies might result in job loss. According to the World Economic Forum

O. Adepoju et al., *Re-skilling Human Resources for Construction 4.0*, Springer Tracts in Civil Engineering, https://doi.org/10.1007/978-3-030-85973-2_9

(2018), there will be approximately 133 million new roles created due to the new division of labour between humans, machines, and algorithms (WEF, 2018). Recently, some workers are now clueless about what to do since machines and robotic devices have replaced their skills. Organizations with such workers should rather retrain them entirely with new technology systems and tools to keep their workforce and keep them vibrant. However, the best possible way to tackle these challenges and addressed the new development in skills is reskilling and upskilling of the human resources. In light of the application of 41R in the construction industry landscape, which involves the use of new technologies, there is a need to emphasise the significance of reskilling and upskilling the workforce in the sector.

9.2 Concept and Significance of Skills

Every nation, organization and industry are made up of people, and its success lies in the skills and competencies of human resources. Human resources, also referred to as human capital in a holistic view, consider skill one of the key elements. Skills are important to the future of any economy, and improving skills is crucial to maintaining the sustainable growth and development of nations. The provision of a skilled workforce promotes creativity, innovation and increased productivity in the various sector of a nation. Skills can produce economic benefits to individuals, groups, organizations, communities and a whole nation. Historically, the concept of skill was described as a manual craft. Over the years, the narrative has changed due to the emergence of new skills as a driver of innovation and competition and economic growth. Skills contribute to creating a greater capacity in workers, which encourages industries to adopt new technologies and ways of working (Productivity Commission, 2017). The world of work is rapidly transforming so that current skills are becoming obsolete, thereby creating more demand for skill upgrades. Skill is considered an important factor that workers ought to adapt to the current changes in various industrial operations ushered by the fourth industrial revolution. Every industry requires certain skills for successful delivery in this fast-changing world influenced by technological innovation. (Adepoju & Aigbavboa, 2020). Consequently, advancement in technology present a greater risk of increasing skill shortage and widening skills gap and mismatch.

The word skill has been defined in several ways based on the field of study. Nevertheless, in general terms, skills are competencies and expertise required to perform a job and conduct the everyday task in the best possible way. It is the delivery of skilled, expert, fast, and accurate behaviours to complete tasks. A skill is the ability of an individual to perform appropriately in every situation. It entails the application of knowledge (explicit and/or tacit), the use of tools, cognitive and practical approaches and procedures, and implies beliefs, dispositions and values (for example, attitudes) (OECD, 2016). Skills can be categorized into two: the hard skills, also referred to as technical, and the soft skills. Both skill categories are important and required from job applicants to possess to be qualified for any job.

9.2.1 Soft Skills

Soft skills are an important requirement for all jobs, which often complement the hard skills. It is called social skills or people skills. It indicates competencies that are not directly associated with a specific task but are essential in any position. They can be described as qualities an employee should possess to be effective on the job. They apply to an individual's required behaviours and attitudes, such as emotional intelligence, which establishes good relationships with people in the work environment. In a corporate organisation, human resource management emphasises soft skills because it is believed to drive a positive working atmosphere that is essential for an organisation's growth. These skills are not necessarily acquired through formal education but an interpersonal relationship in the workplace. Examples of soft skills include communication, interpersonal, teamwork, problem-solving, critical thinking, conflict resolution and negotiation skills. In the current 4th industrial revolution, the human skills under the soft skill which enhance value creation activities for the organisation are of great demand. These skills include emotional intelligence, complex problem solving, critical thinking, analytical thinking, creativity, leadership, innovation and social influence. In the construction industry, soft skills are required to manage a project from the beginning to the end successfully.

9.2.2 Hard Skills

Hard skills are also referred to as technical skills. They can be described as technical expertise, which is an important factor that determines a worker's competence on the job. These skills relate to specific capabilities to accomplish a particular job. It defines what a worker does on his/her job and the level of professionalism. Workers require hard skills to be effective and efficient in any line of job. These skills are significant as it increases workers' efficiency and productivity. Hard skills are learnt and can be acquired through formal and informal education and training programs. Examples of hard skills include computer, digital, programming skills.

9.3 Current Skills Demand for Construction 4.0

Construction 4.0 presents great opportunities for enhancing workers' productiity but also create wide skills gap. This scenario is generating a need for a workforce with a specific set of skills and capabilities. Some schools of thought believe that Construction 4.0 is simply adding to the rising skills gap, other believe that it will create more job opportunities to close the gap. The complexity of construction 4.0 characterized by Cyber-Physical System and Smart Factories, transforming construction production environments, places workers under more pressure to perform well.

The construction work is becoming less routine and demands continuous knowledge and skill development. Workers' skill sets are likely now require better higher-order critical reasoning and decision-making abilities. On the other hand, bridging the skills gap necessitates novel user-facing technology such as Augmented Reality (AR) and wearables for human performance augmentation provided vis live guidance to increase productivity and effectiveness of workers. AR extends beyond mobile applications with 3D object super-imposition for marketing to more complex use cases delivered by an increasingly increasing innovation ecosystem of hardware and software providers working closely with R&D organizations (Kember et al., 2017).

The workforce in the era of Construction 4.0 needs to acquire new skills to update themselves for successful implementation of the new technologies. There is no doubt about it that all the categories of Job will be affected one way or the other. However, professionals working on site will need to acquire new skills as the nature of their job changes. Also, some jobs will be automated, thereby resulting to loss of job. Consequently, for the construction industry to fully enjoy the benefits of construction 4.0, reskilling and upskilling becomes an important focus.

Construction 4.0 is characterized by technologies such as Robotics, Internet of Things, Building Information Modeling, Drones Technology, Prefabrication, 3D Printing. The first step to take in embarking on reskilling and upskilling is the identification of relevant skills. The construction industry will require reskilling in two aspects: technical skills and soft skills. For the purpose of reskilling and upskilling, this section will take a cursory look at the current skills in demand in light of the construction 4.0 technologies.

9.3.1 Building Information Modelling

Building Information Modelling is one of the construction 4.0 technologies that is widely adopted in both developing and developed countries. Building information modeling (BIM) generates and manages digital representations of physical and functional characteristics of places using a variety of tools, technologies and contracts. It demands both the soft and technical skills and the soft skills include: communication skills, interpersonal skills, negotiation and conflict management skills, team playing skills, leadership skills, planning and organizations skills, and the analytical and problem solving skills. On the other hand, the technical skills include: computer literacy skills, information technology and proficiency in BIM software's skills.

9.3.2 Drone/Unmanned Aerial Vehicles (UAVs) Technology

The impact of drones in the construction industry under the construction 4.0 era is very significant. This technology is used more for visualization and video recording. The advancement of UAV technology has necessitated the need for reskilling of workers to effectively operate the technology, push the existing boundaries, and develop new drone applications to benefit the general public. This technology also demands both soft and technical skills. The soft skills include communication, data analytics. The technical skills include programming and machine learning skills.

9.3.3 Prefabrication

Prefabrication is an important manufacturing technology in the Architecture, Engineering and Construction industry that allows building components to be manufactured in a regulated environment away from the construction site. In the era of construction 4.0, prefabrication requires a high level of skills. Consequently, professionals in the construction industry such as architects, engineers, quantity surveyors, and few manual workers such as bricklayers and carpenters are required to possess certain skills. These skills can also be categorized under soft and technical skills. The soft skills include: planning, communication, collaboration, interpersonal, negotiation and team skill. The technical skills include designing, digital, automation, analytical, installation and material handling skills. The designing skill should be a top consideration when reskilling to meet up with the construction 4.0 skill requirements.

9.3.4 3D Printing/Addictive Manufacturing

3D printing, also known as Additive Manufacturing (AM), requires specific skills for effective use in construction 4.0. Recently, new machines and materials are being introduced in the growing and rapidly 3D printing marketplace. Hence, professionals in the industry need to acquire certain skills to utilise the 3D technology in construction 4.0 effectively. These skills can also be categorized under soft and technical skills. The soft skills include critical thinking, problem solving and creativity. The technical skills include software designing modelling, intellectual property protection skill, safety management and 3D post-processing/finishing.

9.3.5 Robotic Technology

Robotics, also known as automation, have contributed tremendously to the improvement in the operations of the construction industry. It is an interdisciplinary field that depends on multiple fields of study. However, there are universal skills professionals should possess for the effective operation of robotic, either as a generalist or a specialist. These skills can be categorised into technical and soft skills. The soft skills include problem-solving, critical thinking, problem-solving, and teamwork.

The technical skill includes computing, programming, data analytics and machine learning.

9.3.6 Internet of Things (IoT)

Internet of Things (IoT) technology is an intelligent technology that presents great potentials to the construction industry. It can connect all devices to the internet and can recognize, configure and transform data to optimize construction operations leading to high productivity, increased business value and cost-effectiveness. IoT applications demand certain skills for effective applications in the construction industry. However, there is a lack of skills and expertise to take advantage of this technology in most organisations, including the construction industry. The skills are primarily hard skills. These skills include Business Intelligence, Data Security, Machine learning, Artificial Intelligence skills, Data Analytics: Teamwork, Designing, Mobile Applications Development, Hardware Interfacing, and Internet Protocol (IP) Networking.

9.4 The Concept of Reskilling and Upskilling in the Fourth Industrial Revolution

The concept of upskilling and reskilling is significant for all industries in the economy to take advantage of the new technologies of the fourth industrial revolution. The two concepts are considered human capital development strategies that are crucial for any organization to remain competitive and be assured of talents required for the future. In the era of the fourth revolution, upskilling and reskilling becomes important to ensure that there are skills available to operate the facilitate the implementation of construction 4.0. It is apparent that any organisation that wants to survive the change brought by the fourth industrial revolution must be given to upskilling and reskilling of their workers. Technology cannot be fought; we can only adapt to its changes. According to the World Economic Forum's (2018), new technologies will displace 75 million jobs across 20 major economies by 2022. However, this does not imply that workers are being priced out of the labour market. According to the same survey, technological advancements would create about 133 million new jobs, which is good news for workers. However, these new positions and experiences would necessitate further training. Companies must develop strategies that will distinguish and make them stand out among their competitors in the markets, both locally and globally. This can be achieved with a flexible, diverse, and creative workforce committed to learning, acquiring and adapting to new skills. Considering the rapid pace of technological change, a scalable and agile workforce must remain ahead of the curve. The culture of Upskilling and reskilling has become a powerful tool that will facilitate the implementation of new technologies and creative business practices. Any proactive industry must take upskilling and reskilling if they desire to stay ahead of the curve. This decision will further demand that organizations conduct a skill gaps analysis to determine the employee who needs to reskill and the skills they need to acquire. In light of this development, the construction industry has been dramatically affected by the fourth industrial revolution, which is referred to as Construction 4.0. This has necessitated the need for reskilling and upskilling of workers. The construction 4.0 technologies present new skills demand which is explored in this book. This has necessitated the need for reskilling and upskilling of the workforce.

9.4.1 Reskilling

Reskilling involves learning new skills other than the current skills required to fit into the modus operandi of doing things. It is the process of preparing an existing employee for a new job or retraining them in the skills needed in today's workplace. This may be a successful technique in a variety of situations (Coonan & Pratt-Adams, 2018). This concept involves acquiring new skills to perform a different job and develop the capacity to train others to perform a different job. The demand for reskilling cut across every category of workers; professionals, skilled, semi-skilled and unskilled. Every individual worker is expected to be dynamic and not get stuck in the old ways of doing things regardless of expertise. Reskilling implies a complete overhaul of skills for employees by taking the workers through the knowledge of new skills entirely within the organisation. It focuses on making employees eligible for other positions within the company. Reskilling creates the opportunity in the organization for lateral learning, making it possible to have people with 'adjacent skills' that are close to the new skills required by the organisation. Reskilling is necessary for the fourth industrial revolution because some skills will become obsolete in the fourth industrial revolution. From a logical point of view, since new technologies are replacing old ones, it follows that most organizations and industries, ranging from the construction industry to other sectors of the economy, might be more tilted towards the new technologies as it guarantees faster methods of operation. Hence, reducing or using old technologies will require workers to learn completely new skills to operate the new technologies of the fourth industrial revolution. Moreover, a company doesn't have to go obsolete before realising that their workers need reskilling; there should be constant evaluation of skills and workflow to see if the present: skills, tools, job-type and functions are still relevant with present technologies.

Once a decline is noticed, then reskilling should be done when necessary. The 4IR is a whole world of opportunities waiting for exploration; thus, learning new skills will surely add to the chance to explore them. The World Economic Forum estimates that the average worker will require 101 days of learning to reskill for the future. Thus, there is a need to train workers on new skills. Reskilling can transcend beyond training or boot camps. It could go as far as enrolling for a new course entirely to get new certifications or attain a new degree. One can decide to go back to school to study a new course relevant in 4IR; thus reskilling is broad and cuts across all forms of learning.

9.4.2 Upskilling

Upskilling follows the same thought of reasoning and purpose as reskilling. Upskilling involves acquiring additional skills to the previous skills acquired in a field. In contrast, an upskilling culture entails teaching employees new, advanced skills to close skill gaps. Upskilling focuses on assisting workers in becoming more professional and relevant in their current roles. This concept occurs when employees enhance their current abilities and have a more significant influence in their expertise. Employees become better positioned for new responsibilities and higher-level tasks on a career path by broadening their expertise. The intention of upskilling is to add more skills already possessed in each area of specialization. The addition of these skills keeps an employee up to date with new technologies as the fourth industrial revolution comes with improved ways of doing things. In this increasingly technology-driven world, a company's capability to upskill, or its capacity for training and developing its personnel to extend their skill sets, generates a substantial competitive edge.

9.4.3 Need for Reskilling and Upskilling in Construction 4.0 Era

Industries under the fourth industrial era, such as the construction industry, present a new working condition characterized by human machine interaction, Big data, interface network problem solving, and control tasks. The construction 4.0 era demands professionals trained in data analytics since most modern technologies are databases that require a vast amount of Big data for operations (Adepoju & Aigbavboa, 2020). According to the OECD (2018), over 1 billion workers, or about one-third of all jobs worldwide, will be transformed by 2030. Some job positions may become obsolete, and at the same time, new ones will be created. As a result, some of the employees' roles might need to be replaced, requiring the industry to either recruit new personnel with the needed skills or engage in reskilling and upskilling of its workforce. Reskilling and Upskilling is necessary for the following reasons:

(i) **Workers' existing skills are becoming redundant**: Digital technology gradually reduces the importance of some workers' skills and areas of expertise in the construction industry. While some construction skills are already obsolete, construction automation is hastening the natural cycle of skills, becoming obsolete at a much faster rate than before. Also, robots are forcing construction workers to change their occupational categories. These will likely lead to some workers becoming redundant and, as a result, laid off. To avoid this occurrence, reskilling and upskilling of workers is required to implement construction 4.0 technologies.

(ii) **Skills Gaps**: Skills gap is one of the challenges faced in the construction industry. Several reasons such as: lack of qualified candidates and skilled professionals, ageing workforce, rising investment in construction projects, dwindling intake into the construction-related profession and lack of necessary skills by employees to meet the changing demand have been attributed to the cause of skills gap. It is believed that the implementation of construction 4.0 technologies will widen and worsen the skills gap. Consequently, upskilling and reskilling becomes a vital strategy to address the problem of skills gap.

(iii) **Accelerated New Technology Adoption**: Today, technology is an integral part of any organisation, which requires that employees possess digital skills. The entry of construction 4.0 comes with new technologies, machines, and techniques present another digital advancement that necessitates digital skills for its integration. Emerging technologies such as Robotics, Artificial Intelligence, Cloud Computing, automation machines and simulation technologies in the construction industry requires special skills to handle them. Hence, upskilling of the workers becomes essential to get them acquainted with the whole idea of construction 4.0. and develop the human capacity for new roles that will emerge.

(iv) **Need for recruitment of new employees**: The construction sector stands a high chance of getting new employees when they commit to training new workers on the new technologies used in construction 4.0. Naturally, candidates who want to feel respected at work will frequently seek out employers with a professional development culture that provides opportunities for reskilling and upskilling.

9.4.4 Importance of Upskilling and Reskilling

(i) **Minimize Skills Gap**: Skills gap is a major challenge that organisations face with introducing new technologies. The swift advancement in technology in the workplace is propelling employers to adopt digital methods of working. The implementation of this is hindered as a result of the skills gap. However, this can be minimized by reskilling and upskilling employees to keep tab of the organisation's current needs. Addressing the skills gap entails the following steps:

- **Awareness**: The industry must be conversant with the area of needs and skills required at and in the future
- **Assessment**: The industry must assess the current workforce to identify areas of skills deficiency.
- **Strategy**: The industry is required to come up with a strategy for upskilling and reskilling the workforce to bridge the skills gap

- **Acceptance**: There must be a level of acceptance on the part of the management and the employees for the need to be committed to the training programs and imbibe the culture of lifelong learning.

(ii) **Reduce Cost of Hiring New Personnel**: Acquiring new personnel to adopt new technology can cost a company a lot. Prospective candidates may demand better compensation because they are applying with specialized skills required and qualifications that are not available among current employees. Also, general recruitment and selection of skilled and competent personnel are expensive and time-consuming. The inclusion of onboarding and training of new employees comes with additional costs. Consequently, reskilling and upskilling current employees will reduce the costs of recruiting and onboarding new employees, reducing the organizational cost of operation.

(iii) **Employee Motivation**: Employees feel highly valued at the workplace when their organisation invests in their development. It is common knowledge that working for a company that does not commit to personal development isn't a great place to be. Employees are motivated when there is the opportunity for active learning, training and developmental programs. This opens the door to professional growth and gives your company a stronger sense of purpose.

(iv) **Constant Updating of Current Skills**: Upskilling of workers will keep workers' skills from becoming obsolete, reflecting the commitment to their professional progress. This process is not restricted to professional training but entails educational processes such as research in the new trends in the industry. Naturally, this will foster innovation and creativity among the staff in the industry.

(v) **Improves Employee Productivity**: Upskilling is an essential element in improving the productivity and efficiency of workers. Through this continuous learning process, workers are motivated to achieve their set targets. Productivity is enhanced through the benefits of flow of communication, multitasking, commitment to strong core values and keeping tap with the new trends offered through upskilling of the employees.

(vi) **Reduce Labour Turnover**: Upskilling is one of the strategies for employee retention. It helps retain employees because organisations that encourage and create learning opportunities have better overall workplace morale. It will make employees feel more at ease in their workplace, but it will also help them be more committed to their jobs and have a more optimistic outlook on their future with the organization. Also, the awareness of the willingness of employers to support their employees in their success help them feel valued builds trust and enhance employee retention.

(vii) **Develop Stronger Organisational Brand and Image**: Reskilling and upskilling practices indicate an organizational commitment to the growth and development of its staff. This propels the workforce to become powerful brand advocates for the organisation. Consequently, it improves an organizational reputation/image and aids in developing a stronger brand in the business market and the society at large.

(viii) **Attract Larger Pool of Skilled Personnel to meet New Technology Skills**: In recruitment, organizations' commitment to reskilling allows them to access a larger talent pool. This opportunity appeal to job seekers who want to enhance their careers and broaden their skill sets. This is also a desirable trait for highly qualified applicants with a wide range of talents and experience. Companies that invest in professional development generate an active learning culture where employees will seek to upskill on their own volition in this climate. Instilling a vibrant learning culture in an organisation will help it be more responsive and prepared the organisation for the ever-changing nature of today's job market.

9.4.5 Strategies for Reskilling and Upskilling for Construction 4.0

Every industry is required to develop an upskilling and reskilling strategy based on their unique need. However, the fourth industrial revolution presents a scenario that affects all industries that specifically entail adopting new technologies. Hence, this requires preparing and implementing effective upskilling and reskilling initiatives. Hence the following approach can be followed to achieved to implement the.

(1) **Assess the Current Situation, Challenges of the Industry**: Upskilling and reskilling the workforce demands assessing the industry's current situation. This can originate from the national industry level or within the company level. The current state of the construction industry and the level of implementation of the industry 4.0 technologies differ from one country to another. Globally, the industry is enthusiastic and gradually integrating automation and digital technologies into its operations. However, security issues such as privacy, data protection, leaked information and risk of data misuse constitute a significant factor limiting the implementation of the technologies (Alaloul et al., 2020). Consequently, the current situation and challenges can be discussed with key stakeholders such as: senior executives of companies, Human Resources leaders, workers' representatives, government officials and representatives from academic and training institutions.

(2) **Assess the Jobs and Skills required: The decision to upskill and reskill requires assessing** the jobs and skills required in the construction 4.0 era. The construction industry is experiencing increasing demand for construction and building projects as a result of urbanization. Across the globe, there is a massive movement of people from rural to urban areas, coupled with demand for better quality projects. This has propelled the desire to adopt the new technologies by the construction companies and professionals to find leverage and make

work more accessible and efficient. Consequently, new jobs are emerging, thereby increasing the skills gap in the industry. Addressing this situation entails assessing the jobs required to meet the industry's increasing demand with the implementation of the new technologies and skills requirements.

(3) **Assess employees' current skills and their capacity to acquire new skills**: The industry, through the service of the Human Resource Managers, is required to assess the employees' current skills to ascertain the skills that are available within the organisation and the competencies that are lacking. This information is needed to organize training to fill the knowledge and skills gaps. It also forms the basis for formulating the reskilling and upskilling plan. The following procedure can be adopted to assess the current skills within an organisation/industry:

- Conduct an examination or test to determine the skills of the employee and the capacity to learn
- The use of personnel evaluation/assessment exercise to determine their current performance and career pro
- Organise a meeting with the employees to get their feedback on the assessment and perception regarding the implementation of the new technologies.
- Identify the skills (soft and hard skills) that are required and employees that are right for upskilling and reskilling

(4) **Develop a Reskilling and Upskilling Plan**: Managers are to develop the reskilling and upskilling plan after identifying the jobs and skills required to implement the new technologies. These plans involve:

(i) Establishing a lifelong learning culture to maximize the success of the upskilling and reskilling plan. This can be achieved through management active involvement and support.
(ii) Design a roadmap to move the employees from their current skill level to the expected skill level. This involves identifying the three pathways of employees which includes employees that require to learn new skills or technology to remain on current job, employees that require reskilling to take up new roles and employees whose job will no longer be required
(iii) Decide on the methods of training, training providers and resources required. The method can include internal and external such as coaching, workshop, seminars, case studies, job rotation, job enlargement, personalized development programs, training/educational courses. Because training requires innovative technologies, updated digital syllabuses should be adopted. It is also essential to choose the best delivery method such as younger employee workers may prefer online while the older may desire traditional training methods
(iv) Design a motivational strategy such as upgrade, incentives, career development to encourage the workers' commitment to the reskilling and upskilling program.

(5) **Execute the Reskilling and Upskilling Plan**: Once the reskilling and upskilling plan are in place, the next step to take is to launch the programs. It involves active participation of managers, employees and human resource personnel. Successful execution of the training plan must ensure adequate training resources such as digital technology facilities and a robust Information Technology system to keep track of the training programs and the employees' attendance. In the case of off-site training, management should support the employees who opt for private career development through sponsorship and the opportunity to participate in the training classes and programs.

(6) **Assess the Success of the Plan**: The purpose of reskilling and upskilling is to meet the organisation's changing needs. Training of the workforce in the light of digitalization of the industry expects skills acquisition and capability to take up new responsibilities and roles. Organisations invest a lot in training and developing their staff. Consequently, assessing training effectiveness is vital as it reveals the impact of the training and informs the management on the area of improvement. It is also advisable that the organisations decide ahead on the measures of training effectiveness. Assessing training effectiveness can be conducted through a feedback mechanism, discussion, oral and practical assessment. In addition, evaluating training effectiveness should be a continuous process to keep the employees updated and well equipped for the future.

9.5 Role of Industry, Academia and Government in Reskilling for Construction 4.0

Skills are fundamental to the development of any economy. Skills required to strive in jobs and the trajectory of careers are swiftly revolving due to technological innovation, significantly changing the labour market's skills (WEF, 2017). The fourth industrial revolution has presented a significant demand for reskilling and upskilling of human capital. This calls for individual and collaborative efforts of the stakeholders. The government (policymakers), industries (business leaders), academia (training and educational institutions) have a role to play in ensuring that training and education systems optimize the availability and skill of the labour force, which demands multi-stakeholder collaboration and investment (WEF, 2017).

The construction industry, like other industries, is experiencing skills mismatch/ gaps and threats of unemployment on the workers with the introduction of 4IR technologies. There is a need to explore the roles of the stakeholders to maximise the impact of improving existing skills (upskilling) and training in new skills (reskilling).

9.5.1 Role of Construction Industry in Reskilling and Upskilling

The construction industry represents an important sector of any given economy, as it is well known to be a catalyst for development alongside agriculture. This sector is responsible for residential and commercial buildings, roads and other social amenities needed for any economy to thrive. Construction 4.0 is gradually transforming the industry with the rise in sustainable buildings constructed faster with less cost and minimize safety risk. However, the demand for this transformation has created a skill gap in the construction sector that requires all the stakeholders with greater responsibility for the industry. The industry must be proactive in ensuring that the modern technologies' skills are in place. At this time, the construction industry must fight to survive by being on the forefront of championing personnel skill development of workers, carrying out skill requirement sensitization, sponsoring training and developmental activities, and partnering with the education institutions. It is believed that the involvement of the industry in reskilling ad upskilling of the workforce for an digitised sector can resolve the problem of the unattractive nature of the construction industry to the young generation. Consequently, the role of industry in reskilling and upskilling is highlighted below:

(i) **Creating Awareness of Skills Demand for Construction 4.0**: The starting point for implementing construction 4.0 is creating awareness of the emerging technologies of the fourth industrial revolution, the benefits and the skills requirement. Every industry, like construction, is taking several steps to ensure the adoption of these new technologies and techniques due to several advantages it presents. However, the pace of adopting these technologies depends on the level of industry commitment in creating awareness and promoting the requirements for the application of the technologies. This can be achieved by industry organizing conferences, developing programs to raise the awareness of critical skills, reskilling and upskilling benefits and training opportunities. Furthermore, the workers and employers in the industry need to be conversant with the essential skills and how to leverage them. This could entail organizing an outreach program to educate workers and employers of existing skills training and support, developing real-time online information and mobile application in the required skills (Asian Development Bank, 2021).

(ii) **Retraining Existing and New Entrants Workforce**: The need for human capacity development has greatly increased in the past two decades due to intense competition and changing business practices. In this present digital age, the concept of reskilling and upskilling isn't just about training. Still, it has grown into a more complex approach of learning and developing not just at the individual level but also at the organizational level (Kareem & Hussein, 2019). In the face of new development, the industry is responsible for training its current workforce and new entrants to ensure that knowledge is updated across ranks and files in the industry. Each organisation is responsible for pushing for reskilling and upskilling its workforce by providing funds for training and development. Also, bonuses and salary upgrades may be used as a driving force to propel its workforce to upgrade his/her skills personally.

(iii) Consequently, the construction industry has a role to play in reskilling and upskilling of the construction workers to get acquitted with the current best practices. The current digital age is changing fast, so organizations should be on the lookout for new technologies to boost their competitive stance. In the past, organizations have used reskilling and upskilling as an essential strategy in stimulating positive individual behaviour, imparting knowledge and skills for optimal and increased productivity and competitiveness in this modern, rapidly changing workforce environment (Kareem & Hussein, 2019).

(iv) **Partnering with Educational Institutions**: Construction firms should take it upon themselves to find the best cost-effective means of reskilling and upskilling its workers. For that to be possible, construction firms need to partner with educational institutions. Partnering with educational institutions constitutes offsite training that allows the current employees to attend courses, acquire knowledge of new introduced technologies, and get certification to upgrade work experience. For example, a construction firm that wants to implement BIM technology can allow an engineer to go for a course in BIM in construction design and get a certification. The partnership with educational institutions would reimagine institutional education the opportunity for students and practitioners to gain knowledge and skills from the revised curriculum in response to the digital era.

(v) **Promoting Work Integrated Learning (WIL)**: Work-integrated learning is vital in meeting the demands of construction 4.0 ushered by the fourth industrial revolution. This scheme is an experiential learning activity that allows a student to apply what is learnt in the workplace. This learning experience allows the learner to acquire the basic work abilities of both soft and technical skills before gainful employment. The challenge faced by most African counties is that industries are not often keen on offering opportunities for a student learning experience. Openings are not available for students to learn WIL program is relevant in reskilling by giving opportunities for

(vi) **Funding Research and Development**: Construction firms themselves can be at the forefront of construction development and investing in research and development, and gaining raw, direct knowledge from field experiences and extensive developmental research. Carrying out research and development in-house would ensure first-hand knowledge update within the organization. Effective research and development would ensure that the right skilling programme is in place, identifying redundant and potential skills and will also help in strategic skill prioritization.

(vii) **Micro-Credentials**: Micro-credentials, also known as industry-recognised certificates, are certifications (often digital) that authenticate or validate a person's proficiency in a definite skill or collection of skills. Micro-credentials have been hailed as a means by which interested persons can skill up in a specific area and bridge the deficiencies conventional Higher Education Institutions (HEI) degrees present. Micro-credentials have the benefit of providing focused training and are a means for employee upskilling and reskilling. These skills can either be hard or soft skills. Another more critical advantage of micro-credentials is that they provide a platform for developing and assessing job competence. The industry has discovered that qualifications do not always translate to competence, and there is the need for alternative methods for competence assessment and development. Construction industries can leverage micro-credentials and offer them to their employees either singly or in conjunction with accredited educational institutions. The goal is to enhance micro-credentialing systems in the construction industry to upskill and develop competence in persons and bridge the skills gap that HEI programmes currently have.

9.5.2 Role of Educational Institutions

Education is still considered a critical factor for promoting all-encompassing economic growth and creating a future of opportunity for all. The educational institutions play a significant role in a wide-ranging reskilling and upskilling agenda. These institutions that comprise the secondary and tertiary lay the foundation for learning and acquiring formal knowledge required in the working environment. The higher institutions specifically offer several courses and train the human capital needed in various industries. These institutions can drive major change by enhancing students' awareness, skills, and competencies to contribute to sustainability, environmental conservation, and other critical objectives. They are responsible for training students to engage critical and responsible people and provide opportunities for lifelong learning to assist them in their social roles (EHEA Rome Communiqué, 2020). Education has advanced based on the needs of society. It serves to prepare the human capital to take on the changing tasks and support the students to develop the skillset required in the future. However, the technologies of the fourth industrial revolution are creating a huge knowledge and skills gap, hence calling for the urgent need for higher education to respond to the demand of upskilling and reskilling of the workforce. In the construction industry, several professionals such as architects, quantity surveying, engineers, and project managers who are a product of the tertiary institutions, lack the skills required to appropriate the new technologies. Construction 4.0 presents technologies such as the Internet of Things, Building Information Modelling (BIM), automation and simulation, data analytics, etc., which demands special skills to operate. Developing these skills requires learning and training, which can be acquired from the educational institutions.

In the upskilling area, higher institutions can design new courses, certificates that will improve the existing knowledge and skillsets of the construction workers. These courses will equip the current workforce to keep their jobs by introducing new technologies and preparing potential workers to take up new jobs. In reskilling, the educational institutions also can develop short- terms courses, programs, and certificates that will help increase the skill level of the current workers. These programs can be coordinated in collaboration with professional and certifying bodies. This avail the workers the opportunity to move from one field to another. Consequently, the role of education institutions in reskilling and upskilling agenda for construction 4.0 include the following:

(i) **Development and Revision of School Curriculum**: A curriculum is a vital tool in the educational system comprised of learning and teaching content for courses offered in schools. It serves as a guide for implementing teaching activities. Developing and reviewing curriculum is an essential process in response to the current ongoing digital revolution that is transforming every field of study. Construction 4.0 can be effectively implemented when smart technologies and techniques, such as data analytics, big data, artificial intelligence, robotics, the internet of things and cybersecurity, has been embedded in the courses taught in educational institutions. The design of the school curriculum, especially that of tertiary institutions, is expected to meet the demands of the labour market. The demands of the recent jobs involve complex problem solving, critical thinking, creativity skills etc., that is required to be developed among the students right from the secondary level. Consequently, there is a need to redesign the school curriculum from secondary level to tertiary to prepare the students for the new jobs and roles evolving as a result of 4IR.

(ii) **Technology-Based Teaching and Learning**: The recent digital revolution is altering every aspect of life, presenting a demand on the educational system to transform the teaching and learning methods. Advancement in technology has provided easy access to information and educational materials which has facilitated the learning process. The fourth industrial revolution, which is characterized by the internet of things, automation, big data, artificial intelligence, requires the adoption of technology-based methods to train and educate students to acquire the necessary skills. Furthermore, educational institutions need to integrate digitalized teaching techniques because the current generation of students are "digitalized natives". This generation is often referred to as Generation Z, access information, communication, and exposure to mobile systems, social networks, and the internet. Different learning styles, self-directed learning, and nanodegrees can be adopted at the tertiary level to promote lifelong learning and meet different learners' categories. The nano degree is designed to equip professionals who desire to learn new innovative skills or advance their existing abilities to work with cutting-edge technology. Also, the implementation of construction 4.0 requires the use of information and technology to train students based on the automated and data-based nature of the technologies. Consequently, in response to need construction 4.0, the educational institution should promptly adopt modern technology to harness their power to train the learners.

(iii) **Collaboration with Industry**: Over the years, higher learning institutions collaborate with industries/companies to ensure quality education and training programs. This collaboration was established to develop human capital that can adapt to the ever-changing business world and emerging technologies. The partnership of the educational institutions with industry is also important to equip the students with employability skills required to be relevant after graduation. The development of construction 4.0 requires skilled professionals that can practically operate and manage the new technologies. Consequently, the skills gap created by construction 4.0 can be managed with the active collaboration of the educational institution with industries for practical experience of the operations of the modern technologies. Furthermore, the higher institutions can contribute to reskilling and upskilling the current workforce through their collaboration with employers. Hence, current required skills can be integrated into the training programs.

9.5.3 Role of Government

The role of government is fundamental to the all-round development of any industry in any given country. Generally, the government is saddled with ensuring that institutions that contribute to economic growth are proactive by formulating and implementing policies. Industry 4.0 presented opportunities and threats resulting in a more open, flexible, dynamic, knowledge and skills-based economy. Embracing this trend would promote a country's global competitiveness, but failure to embrace it may lead to the inability of a country to independently keep public services going and may also lead to reputational damage. Government promptness in adopting new technologies is critical as failure to take advantage of the opportunities it offers might have a long-run effect on the economy. For instance, a nation might experience structural unemployment and gross population crisis in the near future when the government is not proactive in integrating Industry 4.0 technologies. Government policy on industry 4.0 is paying off today in China as companies such as Broad Group benefit from favourable regulations in the country. The introduction of industry 4.0 in the construction industry is being slow down due to a lack of dynamic developmental activities, which require huge capital and heavily skilled manpower requirement. Government has a responsibility to play to bridge the gaps in skills needed to enhance Construction 4.0.

The role of government in the sustainability of both industry 4.0 and construction 4.0 is pivotal to the construction industry and economy at large. The government owes it to both construction firms and industry professionals to provide a favourable socio-economic environment in a workable knowledge-based economy. The role of government intervention in construction 4.0 may be implemented by creating policies that regulate minimum best practices in the industry, policies guiding requirement of knowledge production of institutions of learning and active involvement in communication and extension services when needed. Consequently, the role of the

Government in reskilling and upskilling the agenda for construction 4.0 include the following:

(i) **Policies Regulating Education System to Integrate Emerging Skills of 4IR into the Curriculum**: Institutions of learning occupies a bedrock position in any development agenda, and maintaining standard education heavily rests on the government of a nation. The government can promote upskilling and reskilling towards the acquisition of skills for construction 4.0 through policy formulation to ensure that skills required to implement new technologies are taught in the educational system at any. This can be achieved through curricula review and developing frameworks and standards for skill acquisition. These policies will help address the huge misfit between the labour force and labour force requirement, which in the long run often leads to additional training costs that the construction firms would burn.

(ii) **Provision of Adequate Infrastructure to promote Reskilling and Upskilling**: Every government has the mandate to ensure a digital economy that supports socio-economic growth and development. One of the ways by which government can achieve this mandate is through the provision of adequate infrastructure to ensure an enabling environment towards the adoption of technology and development of human capital. To promote construction 4.0 government can give infrastructural support to leverage the emerging technologies for economic growth. Furthermore, quality education and learning can be enhanced by providing infrastructure and funding for educational institutions. Consequently, government intervention is vital to promote reskilling and upskilling of the human capital through acquisition of technology, promoting research, funding of training institute, providing incentives and sponsorship of human capital locally and internationally

(iii) **Formulate Policies Regulating Minimum Skill Set for Construction Firms**: Policy is one of the strongest mechanisms of the government to regulate systems, implement new standards and combat challenging situations. Government formulating policies on the standard construction practices for Construction 4.0 would help to promote reskilling and upskilling in the construction firms. Diving fully into construction 4.0 without proper regulations might open doors for the high rise of substandard construction activities. The best way to regulate the activities of the industries and maintain standards to facilitate the implementation of the new technologies is to make policies requiring the construction companies to have the necessary expertise working onsite or offsite. Hence, the government has the responsibility through its agencies to ensure that substandard practitioners and unverified professionals are kept off the construction industry to protect consumers and avoid building catastrophe. These policies will pressure the industry and the professionals to commit to reskilling and upskilling to stay relevant on their job and in the industry.

(iv) **Government Partnership with Industries and involvement in Extension and Communication Services**: Government in any nation are positioned to lead development within their jurisdiction. Generally, the government has been known for initiating seminars, workshops, and other sensitization programs to ensure sustainability and improved quality of life. The development of human capital through reskilling and upskilling of the workforce will require government partnership with private organizations such as construction firms to conduct training, seminars, conferences, research and other programs to develop the workforce. On the other hand, the role of extension and communication is to provide critical services support for construction firms in the face of the new technological challenges facing the construction sector due to the activities of industry/construction 4.0. This can be achieved through a government partnership with construction firms and educational institutions in initiating, designing and delivering skill training and development. Also, at this initial state, government intervention would be required to support construction firms with huge capital investment to optimize construction 4.0. Also, the government, through extension and communication, can encourage firms and individuals to take action and work out solutions to the skill challenge rather than waiting for a readymade solution. Since the governments have a wide collection of workforce training programs, the focus should be shifted to the core skills development of individuals and corporations.

9.6 Framework for Reskilling for Construction 4.0

In light of the discussion in previous subsections, there is the need to articulate a high-level framework for reskilling for Construction 4.0. This framework has four key components, as shown in figure, which are discussed below:

Outline Focus: The construction industry is quite vast, and there is the need to narrow down key areas where there is a skill gap in light of emerging construction 4.0 technologies. This entails a comprehensive analysis of the skill gap in each of a company's departments and sections. This could be done using a skill gap analysis and skill requirement analysis, which a firm's human resource department often uses during recruitment. This analysis will ensure the recognition of missing skills and mismatch skills to reskill in line with the technologies of construction 4.0. This will invariably lead to the assessment of construction industry workers and determine those with transferable skills that can fit into construction 4.0. This identification is highly needed to determine the level of skill possessed by the current set of employees and to know the point of entry of construction 4.0 skills. Hence, in achieving this goal, construction firms must focus not just on evaluating their current employees but also on a retrospective evaluation of the intended objectives of previous trainings, development courses, and periodic workers tasks. This will make possible the identification of workers that can be quickly reskilled in line with the needed construction 4.0 technologies. However, the efficiency of the reskilling process depends on the reskilling methodologies and strategies. The methodologies for the reskilling program should be done internally and externally.

The external approach should be made first, and these include: training across industries, development courses and inter-organization cooperation/ collaboration. The internal methodology should focus on assigning related construction 4.0 technologies tasks to the trained employees, micro-credentialing and work-integrated learning (WIL), whereby trained employees have the environment to practice. Furthermore, the organization must implement reskilling strategies that will align construction 4.0 skills with the organization's vision, mission, operational strategies, marketing/ transactional capability, cultural values, and management capability.

Executive Aspects: This portion pertains to the execution of the hitherto defined program for upskilling. Of crucial importance is the use of timelines to ensure the program is timeously and properly implemented. However, organizations must ensure that the reskilling programs are executed in line with these five (5) guidance. Firstly, there must be a periodic reskilling plan for the concerned employees. Also, the execution must be in line with current market reality and industry trends as technological innovations occasioned by construction 4.0 are consistently changing due to research and development (R & D). This will ensure that the reskilling programs do not reinforce or become obsolete with time. Thirdly, the reskilling program must be done with caution to avoid corporate espionage.

Furthermore, the execution must be channel towards enhancing the organisation's profitability by creating new ventures and new market targets. This will bring seriousness and a sense of purpose to the program, ensuring that both senior managers and employees see the program as strategic to the corporate existence of the organization. Lastly, the execution must take cognizance of the attendant risks involved. This can be done by conducting a comprehensive PESTLE analysis, which is the acronym for Political risks, Economic risks, Social risks, Technological risks, Legal risks and Environmental risks.

Evaluation: This portion speaks to the monitoring of the program within predefined timelines to see if the program is fulfilling its intended purpose. Hence, the organization must set specific targets and milestones and practical opportunities to demonstrate what has been learnt. This will also entail a set of supervisors and periodic tests of knowledge acquired by evaluating their level of absorptive capacity of the employees and the organization to measure the progress of construction 4.0.

Communication: This portion speaks to engaging with internal and external stakeholders about the upskilling program and sharing best practices for program improvement and replication. A communicated success story helps standardized reskilling practices into best practices, brings about feedbacks for improvement and through inter-organizational cooperation, replication is possible for the benefits of all.

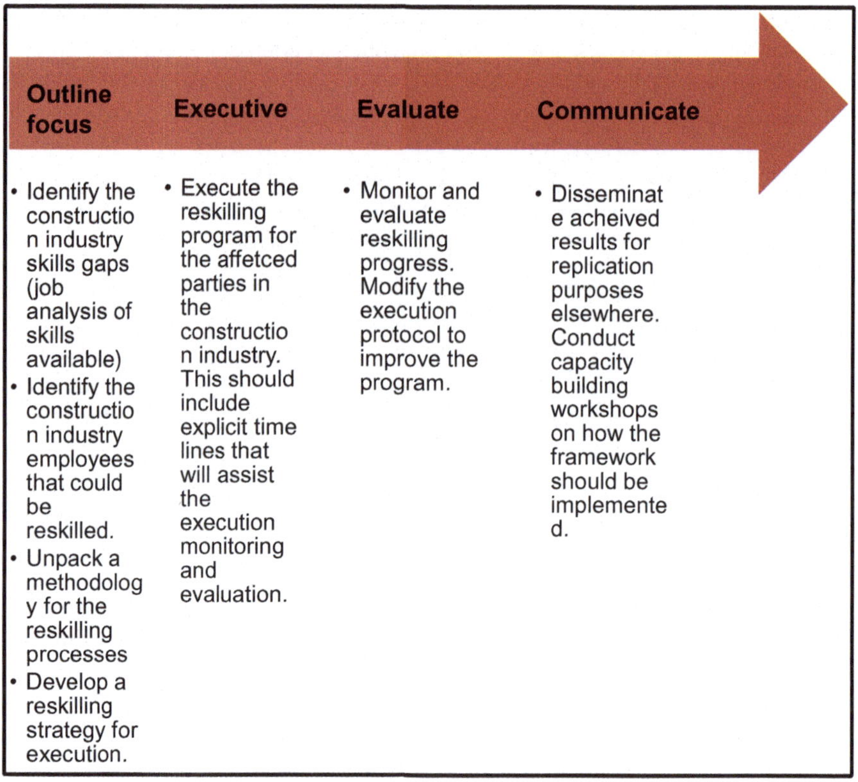

9.7 Summary

Industry 4.0 has kick-started a digital race and has altered and still alters how we live, learn, work, eat and communicate, leaving a huge knowledge gap and expectation. Therefore, to get the most out of construction 4.0, the skills available must match trending technologies. This demands that all key players must be proactive in ensuring that they're not just surviving through the trends but actively gaining through the trends. The expectations to measure up with current tech trends from businesses, employees, and society have taken a new turn. Technology cannot be fought; we can only adapt to its changes. These changes have led to a continuous redefinition of job requirements, creating the necessity to become familiar with the demands of the revolution for quick adaptability. This demand emphasizes the need for human and capital development. To scale through these demands, the involvement and collaborations of industries, government, and educational institutions becomes significant towards developing skills required for this transition. Reskilling and upskilling are considered human capital development strategies that are crucial for any organization to remain competitive and be assured of talents required for the future. Consequently, any organisation that wants to survive the change brought by the fourth industrial revolution must be given to upskilling and reskilling of their workers. On the one hand, workers must possess the needed skills (both soft and hard) to be relevant, for career advancement and sustainability in the industry. Reskilling at any point in time must require both long- and short-term strategies. These strategies must plan to assess: the current situation, challenges of the industry, evaluate the jobs and skills that are required, evaluate and develop a reskilling and upskilling plan, execute the reskilling and upskilling plan, assess the success of the plan, and set feedback strategies at every levy of planning. Currently, the construction industry is experiencing skills mismatch/ gaps and threats of unemployment on the workers with the introduction of 4IR technologies. There is a need to explore the roles of major stakeholders (construction industry, educational institution and the government) to maximise the impact of improving existing skills (upskilling) and training in new skills (reskilling).

References

Adepoju, O. O., & Aigbavboa, C. O. (2020). Assessing knowledge and skills gap for construction 4.0 in a developing Economy. *Journal of Public Affairs*. 1–10. https://doi.org/10.1002/pa.2264.

Alaloul, W. S., Liew, M. S., Want Abdullah Zawawi, N. A., &., Kennedy, B. (2020). Industrial revolution 4.0 in the construction industry: Challenges and opportunities for stakeholders. *Ain Shams Engineering Journal, 11*(1), 225–230. https://doi.org/10.1016/j.asej.2019.08.010.

Asian Development Bank (ADB). (2021). Reaping the Benefits of Industry 4.0 through Skills Development in Vietnam, https://www.adb.org/sites/default/files/publication/671921/benefits-industry-skills-development-viet-nam.pdf

Coonan, E. M., & Pratt-Adams, S. (2018). Building higher education curricula fit for the future: How higher education institutions are responding to the industrial strategy. *The Anglia Ruskin University*. Retrieved from https://arro.anglia.ac.uk/703769/.

EHEA Rome Communiqué (2020). European Higher Education Rome Ministerial Communiqué 19 November, 2020, Retrieved from http://www.ehea.info/Upload/Rome_Ministerial_Communique.pdf.

Kareem, M. A., & Hussein, I. J. (2019). Impact of human resource development on employee performance and organizational effectiveness. *Management Dynamics in the Knowledge Economy, 7*(3), 307–322. https://doi.org/10.25019/MDKE/7.3.02

Kember, D., Hong, C., Yau, V., & Ho, S. (2017). Mechanisms for promoting the development of cognitive, social and affective graduate attributes. *Higher Education, 74*(5), 799–814. https://doi.org/10.1007/s10734-016-0077-x

OECD Organisation for Economic Co-operation and Development. (2016). *The Survey of Adult Skills: Reader's companion* (2nd ed.). OECD Publishing.

Productivity Commission (2017). Shifting the Dial: 5 years productivity review - chapter 3: Future skills and work, viewed 28 October 2017.

World Economic Forum (2017). Accelerating workforce reskilling for the fourth industrial revolution: An agenda for leaders to shape the future of education, gender and work. Retrieved from http://www3.weforum.org/docs/WEF_EGW_White_Paper_Reskilling.pdf.

World Economic Forum (WEF) (2018). The Future of Jobs Report 2018, Retrieved from http://www3.weforum.org/docs/WEF_Future_of_Jobs_2018.pdf